Steve Langfield

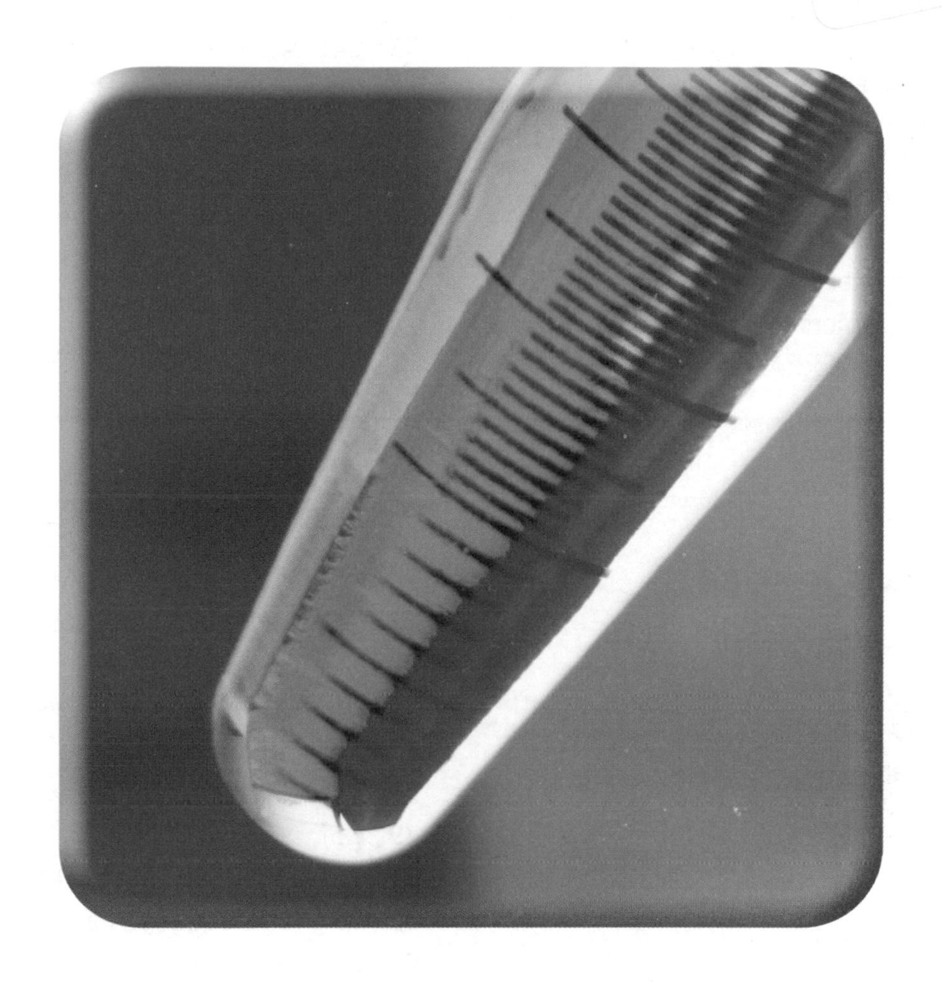

REVISION PLUS

OCR Gateway

GCSE Chemistry

Revision and Classroom Companion

Contents

Fundamental Scientific Processes

Scientists carry out **investigations** and collect **evidence** in order to explain how and why things happen. Scientific knowledge and understanding can lead to the **development of new technologies** that have a huge impact on **society** and the **environment**.

Scientific evidence is often based on data collected through **observations** and **measurements.** To allow scientists to reach conclusions, evidence must be **repeatable, reproducible** and **valid.**

Models

Models are used to explain scientific ideas and the Universe around us. Models can be used to describe:

- a complex idea – like how heat moves through a metal
- a system – like the Earth's structure.

Models make systems or ideas easier to understand by including only the most important parts. They can be used to explain real-world observations or to make predictions. But, because models don't contain all the **variables**, they sometimes make incorrect predictions.

Models and scientific ideas may change as new observations are made and new **data** are collected. Data and observations may be collected from a series of experiments. For example, the accepted model of the structure of the atom has been modified as new evidence has been collected from many experiments.

Hypotheses

Scientific explanations are called **hypotheses** – these are used to explain observations. A hypothesis can be tested by planning experiments and collecting data and evidence. For example, if you pull a metal wire you may observe that it stretches. This can be explained by the scientific idea that the atoms in the metal are arranged in layers that can slide over each other. A hypothesis can be modified as new data is collected, and may even be disproved.

Data

Data can be displayed in **tables**, **pie charts** or **line graphs.** In your exam you may be asked to:

- choose the most appropriate method for displaying data
- identify trends
- use data mathematically – including using statistical methods, calculating the mean and calculating gradients of graphs.

A Table

% Yield / Pressure	Temperature			
	250°C	350°C	450°C	550°C
200 atm	73%	50%	28%	13%
400 atm	77%	65%	45%	26%

A Pie Chart

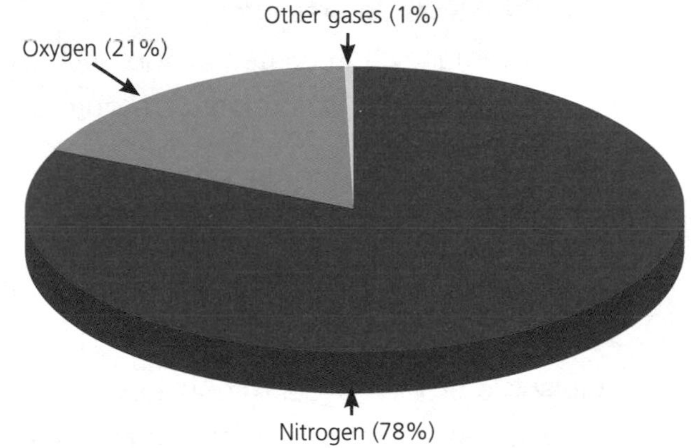

A Line Graph

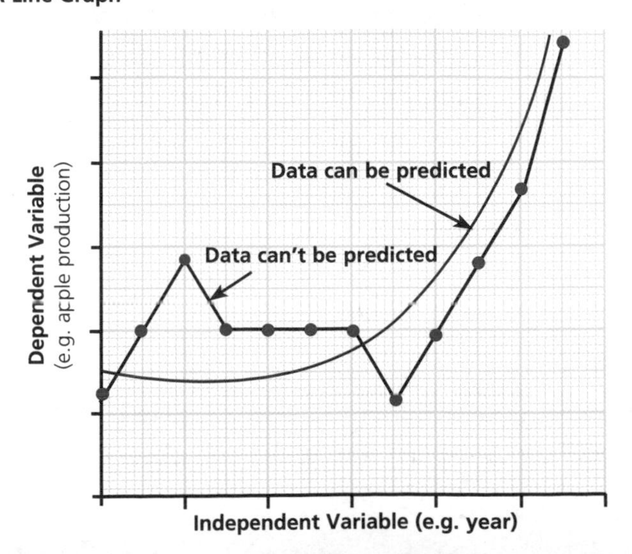

Data (cont)

Sometimes the same data can lead to different conclusions. For example, data shows that the world's average temperatures have been rising significantly over the last 200 years. Some scientists think this is due to increased combustion of fossil fuels, while other scientists think it's a natural change that has happened before during Earth's history.

Scientific and Technological Development

Every scientific or technological development can have effects that we do not know about. This can give rise to **issues**. An issue is an important question that is in dispute and needs to be settled. Issues can be:

- **social** – they impact on the human population of a community, city, country or the world
- **environmental** – they impact on the planet, its natural ecosystems and resources
- **economic** – money and related factors such as employment and the distribution of resources
- **ethical** – what is right and wrong morally; a value judgement must be made
- **cultural** – giving an insight into differences between people on local and global scales.

Peer review is a process of self-regulation involving experts in a particular field who **critically examine** the work undertaken. Peer review methods are designed to maintain standards and provide **credibility** for the work that has been carried out. The methods used vary depending on the nature of the work and also on the overall purpose behind the review process.

Evaluating Information

Conclusions can then be made based on the scientific evidence that has been collected – they should try to explain the results and observations.

Evaluations look at the whole investigation. It is important to be able to evaluate information relating to social–scientific issues. **When evaluating information:**

- make a list of **pluses** (pros)
- make a list of **minuses** (cons)
- consider how each point might **impact on society**.

You also need to consider if the source of information is reliable and credible and to consider opinions, bias and weight of evidence.

Opinions are personal viewpoints – those backed up by valid and reliable evidence carry far more weight than those based on non-scientific ideas. Opinions of experts can also carry more weight than those of non-experts. Information is **biased** if it favours one particular viewpoint without providing a balanced account. Biased information might include incomplete evidence or it might try to influence how you interpret the evidence.

Examples of these processes are included within the main content of the book. However, it is important to remember that fundamental scientific processes are relevant to all areas of science.

Fundamental Chemical

You need to have a good understanding of the concepts (ideas) on the next four pages, so make sure you revise this section before each exam.

Elements and Compounds

An **element** is a substance made up of just one type of **atom**. Each element is represented by a different chemical symbol, for example:

- Fe represents iron
- Na represents sodium.

Atoms have a positive nucleus orbited by negative electrons.

These elements (and their chemical symbols) are all arranged in the **Periodic Table** (see below).

Compounds are substances formed from the atoms of two or more elements, which have been joined together by one or more chemical bonds, for example, H_2O, $CaCO_3$ and $C_6H_{12}O_6$.

Ions are atoms or small molecules that have a charge, for example, Na^+, Cl^-, NH_4^+ and SO_4^{2-}.

A positive ion is formed when an atom loses electrons; a negative ion is formed when an atom gains electrons.

Covalent bonds are formed when two atoms share a pair of electrons. (The atoms in molecules are held together by covalent bonds.)

Ionic bonds are formed when atoms lose or gain electrons to become charged ions; the positive ions attract the negative ions.

The Periodic Table

1	2											3	4	5	6	7	0
								1 **H** hydrogen 1									4 **He** helium 2
7 **Li** lithium 3	9 **Be** beryllium 4											11 **B** boron 5	12 **C** carbon 6	14 **N** nitrogen 7	16 **O** oxygen 8	19 **F** fluorine 9	20 **Ne** neon 10
23 **Na** sodium 11	24 **Mg** magnesium 12											27 **Al** aluminium 13	28 **Si** silicon 14	31 **P** phosphorus 15	32 **S** sulfur 16	35.5 **Cl** chlorine 17	40 **Ar** argon 18
39 **K** potassium 19	40 **Ca** calcium 20	45 **Sc** scandium 21	48 **Ti** titanium 22	51 **V** vanadium 23	52 **Cr** chromium 24	55 **Mn** manganese 25	56 **Fe** iron 26	59 **Co** cobalt 27	59 **Ni** nickel 28	63.5 **Cu** copper 29	65 **Zn** zinc 30	70 **Ga** gallium 31	73 **Ge** germanium 32	75 **As** arsenic 33	79 **Se** selenium 34	80 **Br** bromine 35	84 **Kr** krypton 36
85 **Rb** rubidium 37	88 **Sr** strontium 38	89 **Y** yttrium 39	91 **Zr** zirconium 40	93 **Nb** niobium 41	96 **Mo** molybdenum 42	[98] **Tc** technetium 43	101 **Ru** ruthenium 44	103 **Rh** rhodium 45	106 **Pd** palladium 46	108 **Ag** silver 47	112 **Cd** cadmium 48	115 **In** indium 49	119 **Sn** tin 50	122 **Sb** antimony 51	128 **Te** tellurium 52	127 **I** iodine 53	131 **Xe** xenon 54
133 **Cs** caesium 55	137 **Ba** barium 56	139 **La*** lanthanum 57	178 **Hf** hafnium 72	181 **Ta** tantalum 73	184 **W** tungsten 74	186 **Re** rhenium 75	190 **Os** osmium 76	192 **Ir** iridium 77	195 **Pt** platinum 78	197 **Au** gold 79	201 **Hg** mercury 80	204 **Tl** thallium 81	207 **Pb** lead 82	209 **Bi** bismuth 83	[209] **Po** polonium 84	[210] **At** astatine 85	[222] **Rn** radon 86
[223] **Fr** francium 87	[226] **Ra** radium 88	[227] **Ac*** actinium 89	[261] **Rf** rutherfordium 104	[262] **Db** dubnium 105	[266] **Sg** seaborgium 88	[264] **Bh** bohrium 107	[277] **Hs** hassium 108	[268] **Mt** meitnerium 109	[271] **Ds** darmstadtium 110	[272] **Rg** roentgenium 111							

Chemical Concepts

… with numbers to write … he composition of compounds. Formul… used to show:

- the different elements in a compound
- the number of atoms of each element in the compound
- the total number of atoms in the compound.

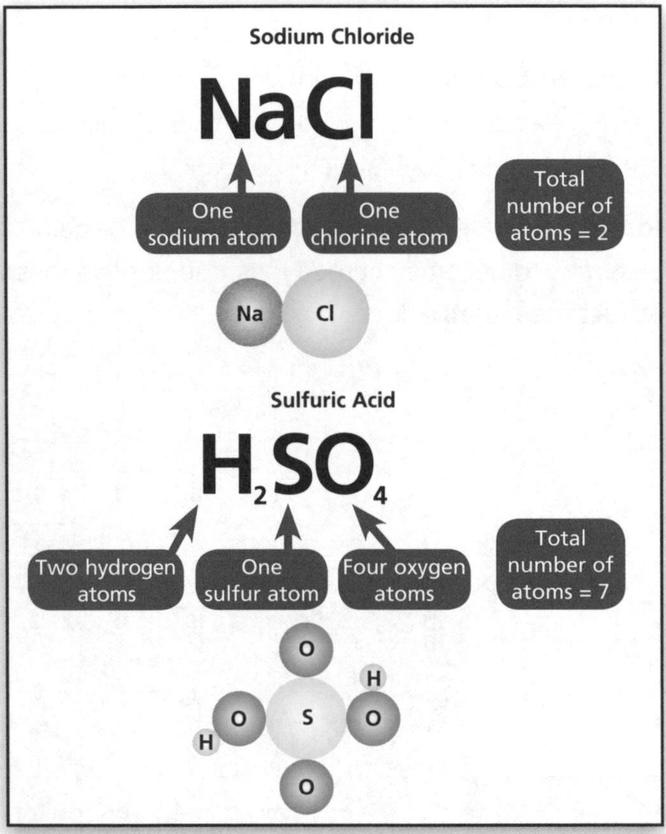

If there are brackets around part of the formula, everything inside the brackets is multiplied by the number outside the bracket.

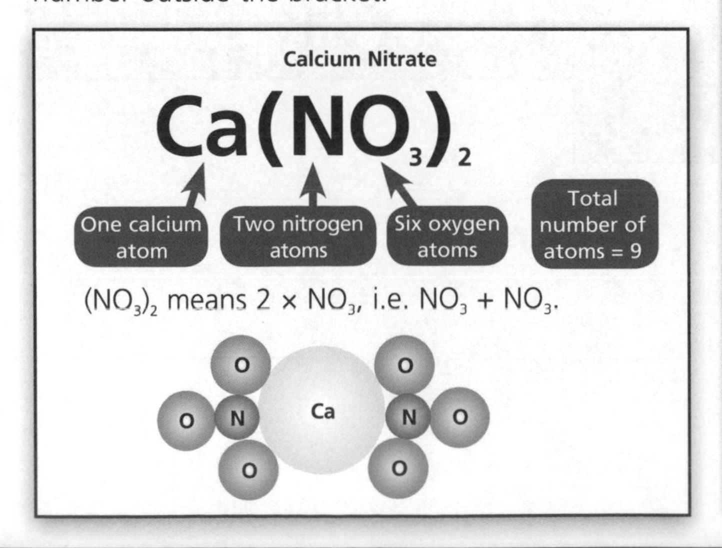

Displayed Formulae

A **displayed formula** is another way to show the composition of a molecule.

A displayed formula shows:

- the different types of atom in the molecule, e.g. carbon, hydrogen
- the number of each different type of atom
- the covalent bonds between the atoms.

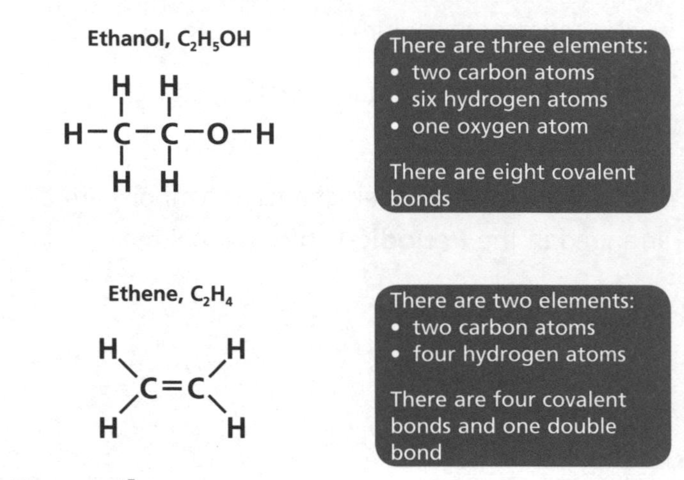

Equations

In a chemical reaction, the substances that you start with are called **reactants**. During the reaction, the atoms in the reactants are rearranged to form new substances called **products**.

Chemists use equations to show what has happened during a chemical reaction. The reactants are on the left side of the equation, and the products are on the right.

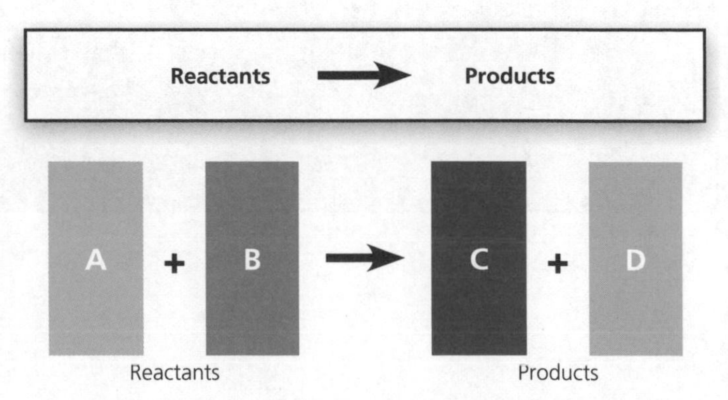

No atoms are lost or gained during a chemical reaction so equations must be **balanced**: there must always be the same number of atoms of each element on both sides of the equation.

Writing Balanced Equations

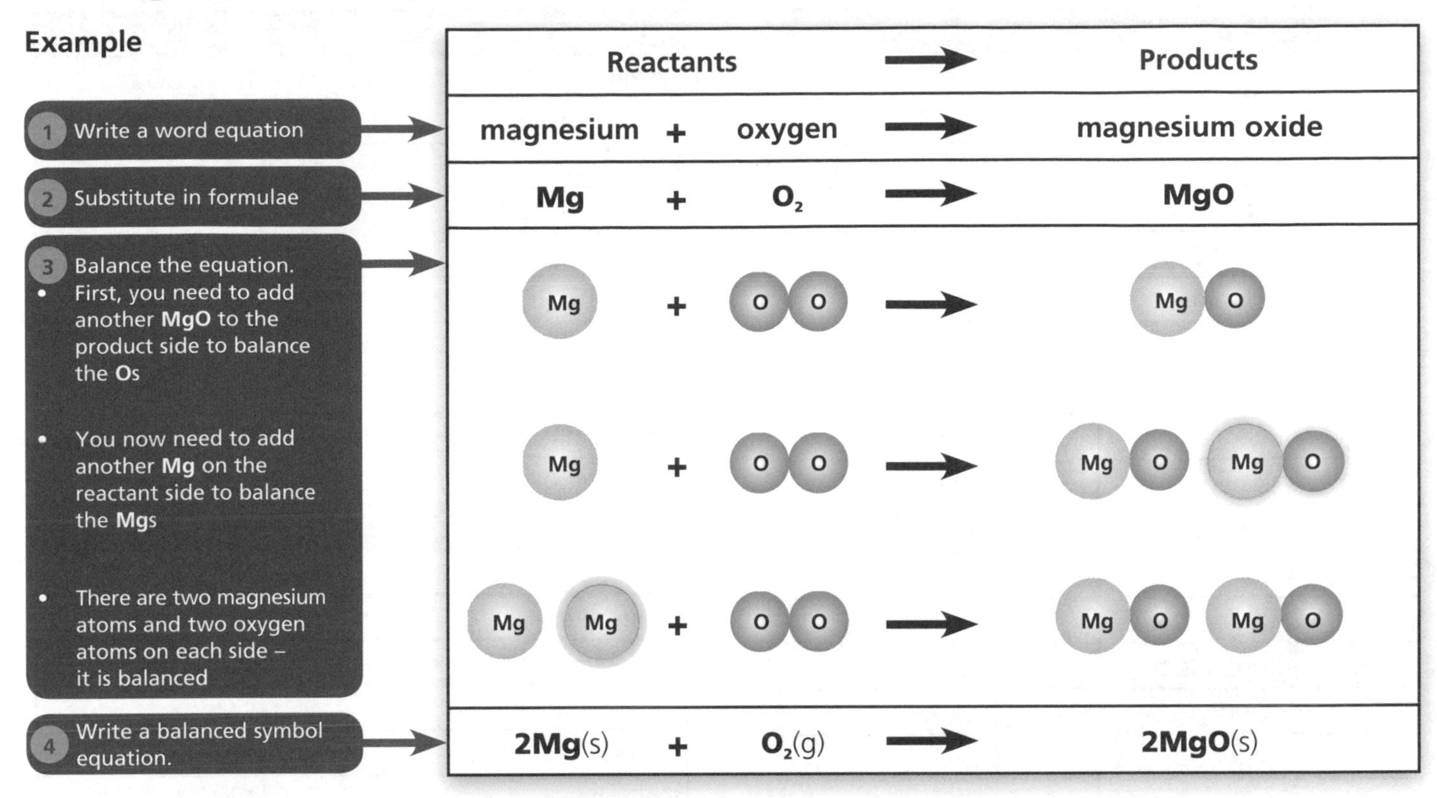

You may be asked to include the **state symbols** when writing an equation: (aq) for aqueous solutions, (g) for gases, (l) for liquids and (s) for solids.

HT You should be able to balance equations by looking at the formulae (i.e. without drawing the atoms).

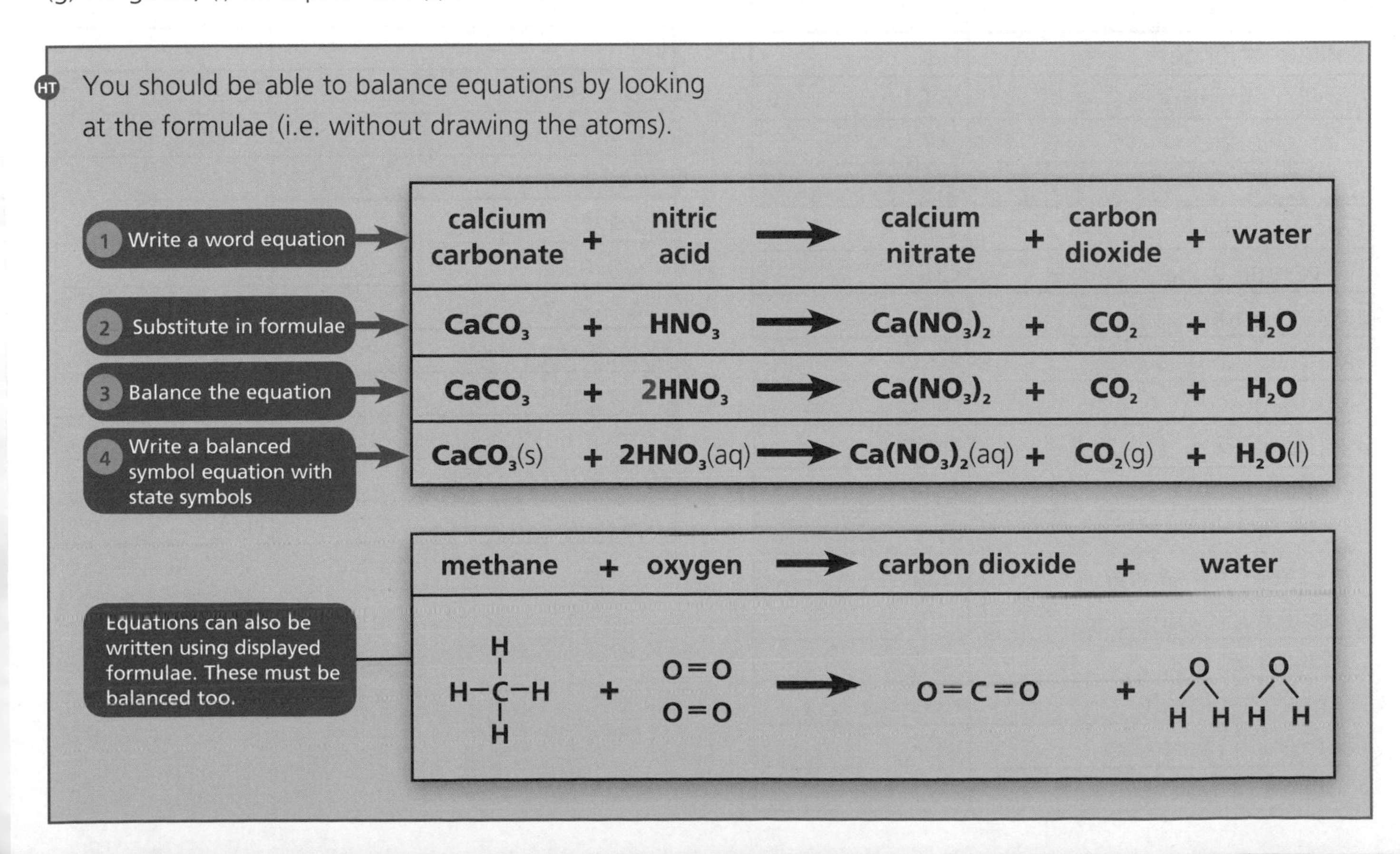

Fundamental Chemical Concepts

Common Compounds and their Formulae

Acids	
Ethanoic acid	CH_3COOH
Hydrochloric acid	HCl
HT Nitric acid	HNO_3
HT Sulfuric acid	H_2SO_4
Carbonates	
Calcium carbonate	$CaCO_3$
HT Copper(II) carbonate	$CuCO_3$
HT Iron(II) carbonate	$FeCO_3$
HT Magnesium carbonate	$MgCO_3$
HT Manganese carbonate	$MnCO_3$
HT Sodium carbonate	Na_2CO_3
HT Zinc carbonate	$ZnCO_3$
Chlorides	
HT Ammonium chloride	NH_4Cl
HT Barium chloride	$BaCl_2$
HT Calcium chloride	$CaCl_2$
HT Iron(II) chloride	$FeCl_2$
HT Magnesium chloride	$MgCl_2$
Potassium chloride	KCl
HT Silver chloride	$AgCl$
Sodium chloride	$NaCl$
HT Tin(II) chloride	$SnCl_2$
HT Zinc chloride	$ZnCl_2$
Oxides	
Calcium oxide	CaO
HT Copper(II) oxide	CuO
HT Iron(II) oxide	FeO
HT Magnesium oxide	MgO
HT Manganese(II) oxide	MnO
HT Sodium oxide	Na_2O
HT Zinc oxide	ZnO
Hydroxides	
HT Copper(II) hydroxide	$Cu(OH)_2$
HT Iron(II) hydroxide	$Fe(OH)_2$
HT Iron(III) hydroxide	$Fe(OH)_3$
HT Lithium hydroxide	$LiOH$
HT Potassium hydroxide	KOH
HT Sodium hydroxide	$NaOH$

Sulfates	
HT Ammonium sulfate	$(NH_4)_2SO_4$
HT Barium sulfate	$BaSO_4$
HT Calcium sulfate	$CaSO_4$
HT Copper(II) sulfate	$CuSO_4$
HT Iron(II) sulfate	$FeSO_4$
HT Magnesium sulfate	$MgSO_4$
HT Potassium sulfate	K_2SO_4
HT Sodium sulfate	Na_2SO_4
HT Tin(II) sulfate	$SnSO_4$
HT Zinc sulfate	$ZnSO_4$
Others	
Ammonia	NH_3
Bromine	Br_2
HT Calcium hydrogencarbonate	$Ca(HCO_3)_2$
Carbon dioxide	CO_2
Carbon monoxide	CO
Chlorine	Cl_2
HT Ethanol	C_2H_5OH
HT Glucose	$C_6H_{12}O_6$
Hydrogen	H_2
Iodine	I_2
HT Lead iodide	PbI_2
HT Lead(II) nitrate	$Pb(NO_3)_2$
HT Methane	CH_4
Nitrogen	N_2
Oxygen	O_2
HT Potassium iodide	KI
HT Potassium nitrate	KNO_3
HT Silver nitrate	$AgNO_3$
HT Sodium hydrogencarbonate	$NaHCO_3$
HT Sulfur dioxide	SO_2
Water	H_2O

C1: Carbon Chemistry

This module looks at:

- How crude oil is processed and used, and the problems with its exploitation.
- Fuels and complete and incomplete combustion.
- The air and pollution problems caused by burning fuels.
- Alkanes, alkenes and how alkenes can be made into polymers.
- The properties, uses and disposal of polymers.
- Chemical reactions in cooking, and the uses of food additives.
- Natural and synthetic perfumes, esters and solutions.
- How paints are made and the use of thermochromic and phosphorescent pigments.

Fossil Fuels

Crude oil, **coal** and **natural gas** are all **fossil fuels**. Fossil fuels are formed extremely slowly. It takes a very long time: millions of years! All fossil fuels are **finite**, i.e. there are limited supplies. They are described as **non-renewable** because we are using them up much faster than new supplies can be formed. This means they will eventually run out.

HT All the crude oil that is easily extracted will eventually run out and any remaining new supplies will have to be found. The search will be in more remote parts of the world and the extraction will be increasingly difficult. As oil becomes scarce, then the decision will have to be made whether to burn it as a fuel or to make petrochemicals (e.g. plastics).

Crude Oil

Crude oil is found trapped in the Earth's crust. To release the oil, a hole is drilled through the rock. If the oil is under pressure, it will flow out. If it is not under pressure, it has to be pumped out.

When crude oil is extracted it is a **thick**, **black**, **sticky liquid**. It is transported to a **refinery** through a pipeline or in oil tankers. This is a dangerous procedure: if the oil accidentally spills into the sea, it can have a devastating effect on wildlife. Oil spills or **slicks** float on the sea's surface. The toxic oil can coat the feathers of sea birds preventing them from floating or flying and may kill them. If an oil slick washes ashore it can damage the beaches leading to large clean-up operations. Detergents are often used to disperse oil slicks or remove oil from beaches but detergents are also toxic to wildlife.

Fractional Distillation

Crude oil is a mixture of many **hydrocarbons**. A hydrocarbon is a molecule that contains only **carbon** and **hydrogen** atoms. Different hydrocarbons have different boiling points. This means crude oil can be separated into useful products or **fractions** (parts) by **heating** it in a process called **fractional distillation**.

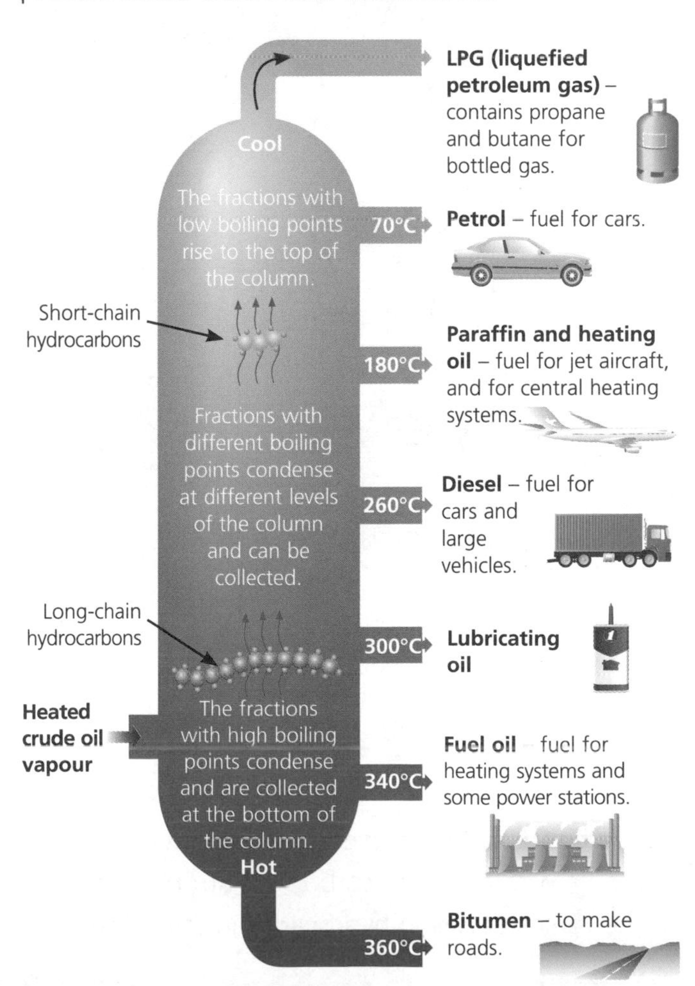

Forces Between Molecules

The **atoms** in a hydrocarbon molecule are strongly held together by the bonds between them, for example:

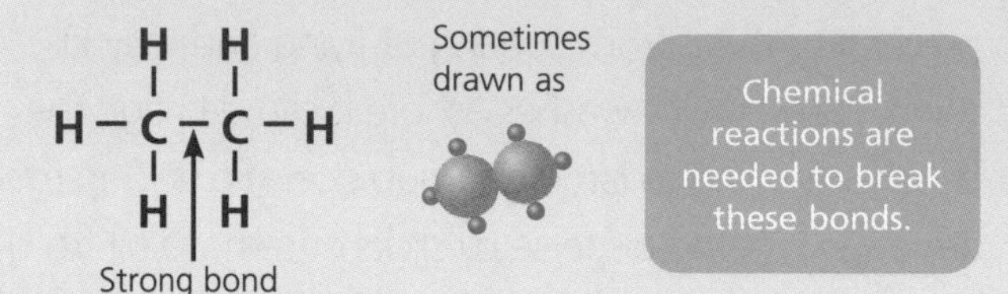

All hydrocarbon molecules have forces of attraction between them called **intermolecular forces**, but they are only weak. However, the longer the hydrocarbon molecule is, the stronger the intermolecular forces are, for example:

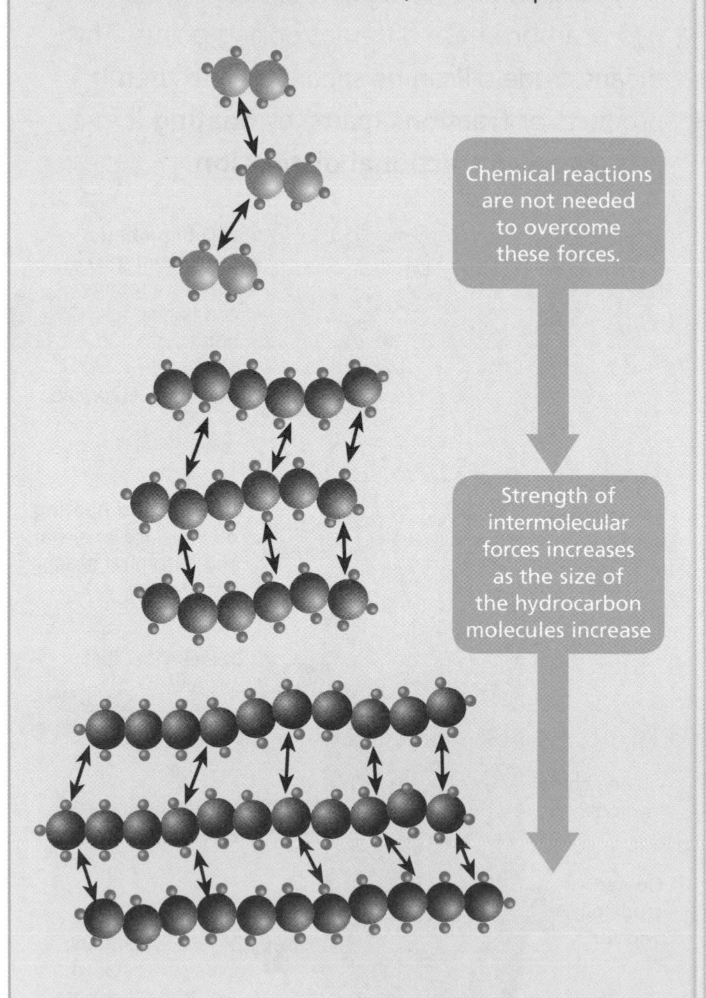

The bonds between the carbon and hydrogen atoms within a hydrocarbon are stronger than the forces between hydrocarbon molecules.

Separating Hydrocarbons

When a hydrocarbon is heated, its molecules move faster and faster until the intermolecular forces are broken.

Small molecules have very small forces of attraction between them and are easy to break by heating. This means that hydrocarbons with small molecules are volatile liquids or gases with low boiling points, for example:

- Methane, CH_4, has a boiling point of -164°C.
- Ethane, C_2H_6, has a boiling point of -89°C.

Short-chain Hydrocarbons

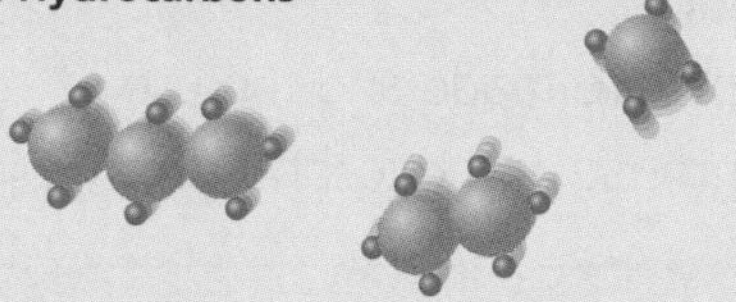

Large molecules have many more of these small forces between them, resulting in an overall large force of attraction. This force is more difficult to break by heating and hydrocarbons with large molecules are thick, viscous liquids or waxy solids with higher boiling points, for example:

- Octane, C_8H_{18} has a boiling point of 126°C.
- Decane, $C_{10}H_{22}$ has a boiling point of 174°C.

Long-chain Hydrocarbons

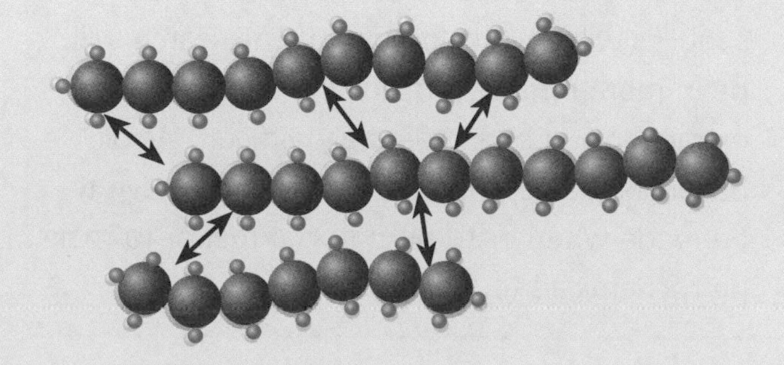

It is the differences in their boiling points which enables a mix of hydrocarbons (e.g. crude oil) to be separated by the process of fractional distillation.

Making Crude Oil Useful

The Fractions

Each fraction consists of a mixture of hydrocarbons whose boiling points fall within a particular **range**.

This table shows the main fractions obtained through the industrial fractional distillation of crude oil, and their approximate boiling ranges.

Fraction	Boiling Range
Refinery gases	up to 25°C
Petrol	40–100°C
Paraffin and heating oil	150–250°C
Diesel	220–350°C
Lubricating oil	over 350°C
Fuel oil	over 400°C
Bitumen	over 400°C

Cracking

Hydrocarbon molecules can be described as **alkanes** or **alkenes**, depending on whether or not they have a carbon–carbon double bond present (see page 14).

Cracking converts **large** alkane molecules into **smaller**, more useful, alkane and alkene molecules. The alkene molecules obtained can be used to make polymers, which have many uses (see pages 16–17). The smaller alkane molecules obtained are usually blended to make petrol, which is in huge demand. You may be asked to answer questions about the supply and demand of crude oil fractions in the exam – the information will be given to you to interpret.

To take place, cracking needs a **catalyst** and a **high temperature**. In the laboratory, cracking is carried out using the apparatus shown below.

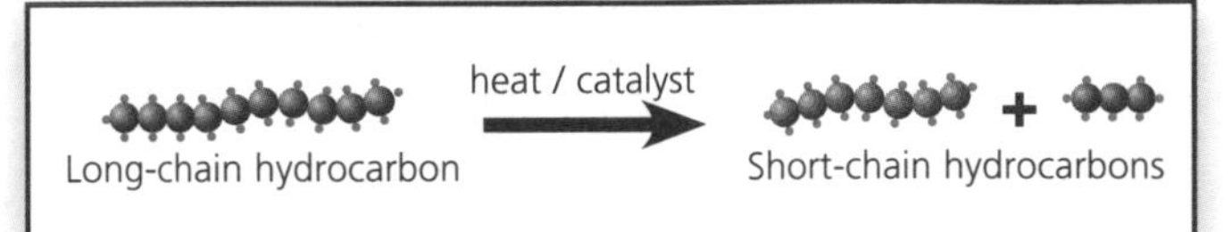

HT There is not enough petrol in crude oil to meet demand. Therefore, cracking is used to convert parts of crude oil that cannot be used into additional petrol.

Crude oil and natural gas are found in many parts of the world. The UK is now dependent on oil and gas from some politically unstable countries, which could cause supply problems in the future.

A Cracking Reaction

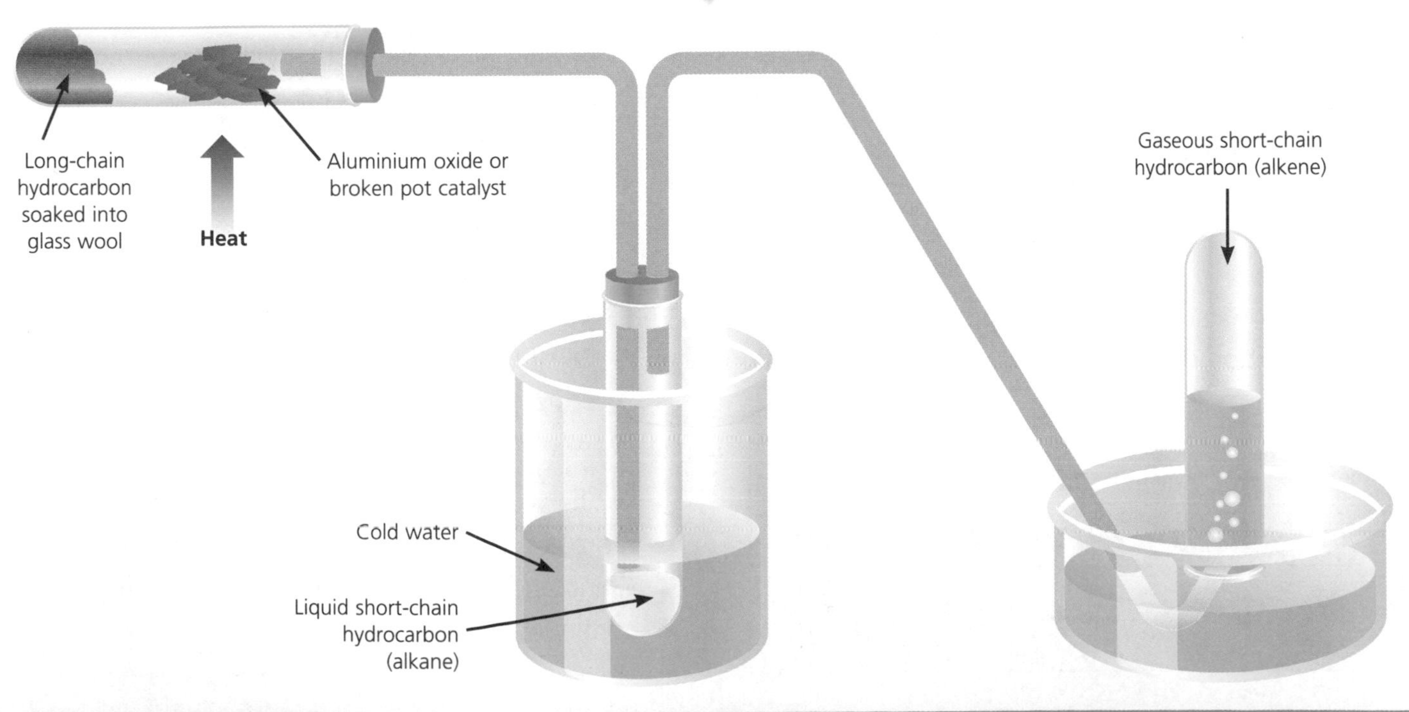

Choosing a Fuel

Some or all of the following factors should be taken into account when choosing a fuel for a specific purpose:

- **Energy value** – how much energy do you get from a measured amount of fuel?
- **Availability** – is the fuel easy to obtain?
- **Storage** – how easy is it to store the fuel? (e.g. petrol is more difficult to store than coal.)
- **Cost** – how much fuel do you get for your money?
- **Toxicity** – is the fuel (or its combustion products) poisonous?
- **Pollution** – do the combustion products pollute the atmosphere? (e.g. acid rain or the greenhouse effect)
- **Ease of use** – is it easy to control and is special equipment needed?

You may be asked to look at data about a number of fuels and decide which one is the best for a particular purpose.

Burning Fuels (Combustion)

When fuels burn, useful energy is released as heat. Chemists call this **combustion**. Fuels are substances that react with oxygen in the air. Complete combustion needs a plentiful supply of oxygen.

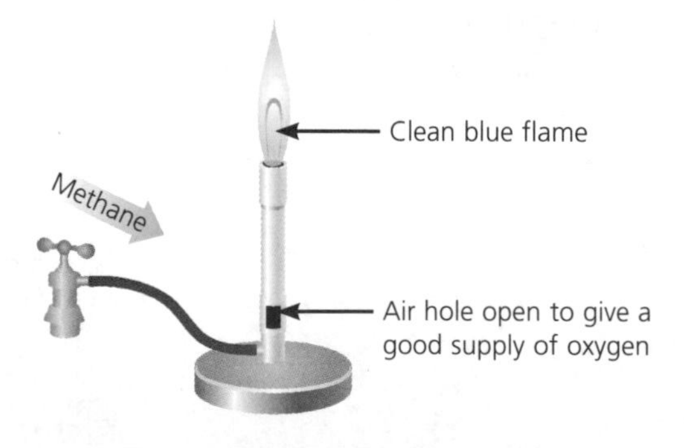

When a hydrocarbon, like methane, is burned in air, only carbon dioxide and water (hydrogen oxide) are formed.

methane	+	oxygen	⟶	carbon dioxide	+	water
CH_4	+	$2O_2$	⟶	CO_2	+	$2H_2O$

Evaluating a Fuel

Choosing a fuel to use for a particular job requires a careful study of available information.

In your exam you may be asked to evaluate the use of different fossil fuels using given data, e.g. tables, graphs, pie charts. The following are examples of the sort of fuels that might be considered.

Methane (CH_4)

- Colourless gas.
- Burns to form carbon dioxide and water.
- Non-toxic (but it is a greenhouse gas).
- Readily available through normal gas supplies.
- Not easy to store.
- 1g of methane produces 55.6kJ of energy when completely burned.

Butane (C_4H_{10})

- Easier to store and transport than methane.
- Burns in the same way as methane.
- Used as camping gas.
- Only 26.9kJ of energy is produced from 1 gram when it is burned.

Coal

- Easy to store.
- Readily available, not very expensive and releases quite a lot of energy when burned.
- Main problem is pollution and in particular the sulfur dioxide gas (that leads to acid rain) it produces, along with smoke and other pollutants.
- Most major populated areas of the UK allow only smokeless coal to be burned as a fuel.

As the world's population increases and more countries become industrialised (e.g. China and India), the demand for fossil fuels continues to grow.

Detecting the Products of Combustion

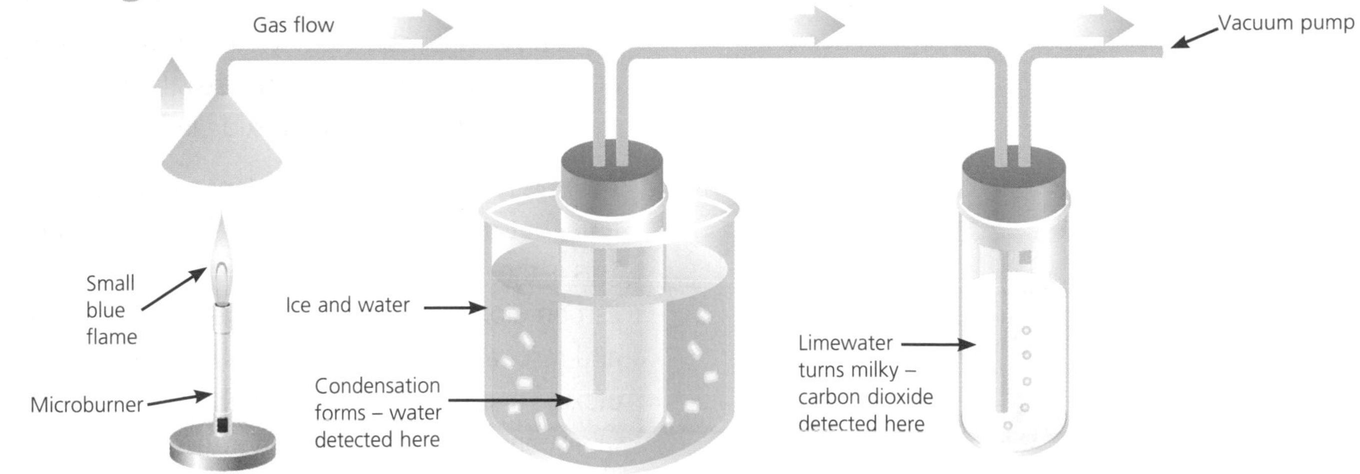

Incomplete Combustion

If a fuel burns without sufficient oxygen, e.g. in a room with poor ventilation or when a gas appliance needs servicing, then **incomplete combustion** takes place and **carbon monoxide** (a poisonous gas) can be formed. For example, the incomplete combustion of methane:

methane	+	oxygen	→	carbon monoxide	+	water
HT $2CH_4$	+	$3O_2$	→	$2CO$	+	$4H_2O$

If there is **very little oxygen** available, **carbon** (soot) is produced instead. For example, the burning of methane when very little oxygen is available:

methane	+	oxygen	→	carbon	+	water
HT CH_4	+	O_2	→	C	+	$2H_2O$

Although incomplete combustion releases some energy, much more is released when complete combustion takes place. The following are also advantages of making sure a fuel burns completely:

- Less soot is produced.
- No poisonous carbon monoxide gas is produced.

A blue flame on a Bunsen burner transfers more energy than a yellow flame because it involves complete combustion. The yellow flame shows that incomplete combustion is taking place.

You should be able to write the word equations for complete and incomplete combustion if you are given the name of the fuel. You may not be given all the names of some of the reactants and products so remember the following points:

- Combustion is a reaction with oxygen.
- Complete combustion of a hydrocarbon produces water and carbon dioxide.
- Incomplete combustion of a hydrocarbon produces water and carbon monoxide.
- Incomplete combustion when there is only a small amount of oxygen produces water and carbon.

HT You should be able to write a balanced symbol equation for complete and incomplete combustion given the formula of the fuel. Remember:

- The formula of oxygen is O_2.

C1 Clean Air

The Changing Atmosphere

The Earth's atmosphere has not always been the same as it is today. It has gradually changed over billions of years.

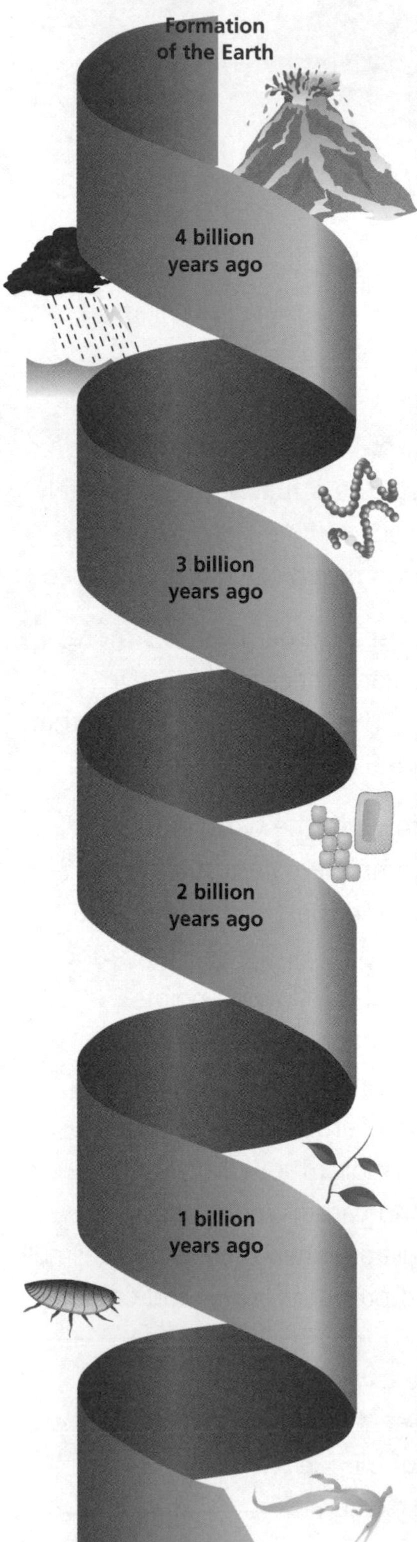

The earliest atmosphere contained ammonia and carbon dioxide. These gases came from inside the Earth and were often released through the action of volcanoes.

As the temperature of the planet fell, the water vapour in the atmosphere condensed to form the oceans and seas.

The evolution of plants meant that photosynthesis started to reduce the amount of carbon dioxide and increase the amount of oxygen in the atmosphere.

Clean air contains about:

- **78% nitrogen**
- **21% oxygen**
- **1% other gases, including 0.035% carbon dioxide**.

It also contains varying amounts of **water vapour** and some **polluting gases**.

Respiration and Photosynthesis

All living things respire. They take in oxygen and give out carbon dioxide. This decreases the oxygen levels and increases the carbon dioxide levels in the air.

Animals respire all the time; plants also respire all the time, but during the day plants also photosynthesise. This is the opposite of respiration: they take in carbon dioxide and release oxygen. Photosynthesis and respiration balance out, so the levels of carbon dioxide and oxygen in the present-day air remain fairly constant. The levels of nitrogen in the air also stay fairly constant.

HT Theories of Atmospheric Change

Explanations about how our planet and its atmosphere have evolved are scientists' best efforts at interpreting all the available evidence. Below is one suggested explanation. But remember, it is only a theory!

A hot volcanic Earth would have released various gases into the atmosphere, just as volcanoes do nowadays. These gases would probably have included water vapour and carbon dioxide (and small amounts of ammonia, methane and sulfur dioxide).

As the Earth cooled down, its surface temperature would have gradually fallen below 100°C and water vapour would have condensed to form the oceans.

The levels of carbon dioxide started to decrease as it dissolved in the oceans. Further reduction in carbon dioxide levels came about when algae and simple plants evolved and used it for photosynthesis. This process also led to the increase in oxygen levels in the atmosphere.

Ammonia in the early atmosphere was converted into nitrogen by the action of bacteria. Nitrogen is very unreactive and so the level of nitrogen in the air has gradually increased.

Air Pollution

Pollutant gases are formed from:

- the burning of fossil fuels
- incomplete combustion in a car engine.

In

- Air – mostly nitrogen and oxygen.
- Hydrocarbon fuel such as petrol or diesel.

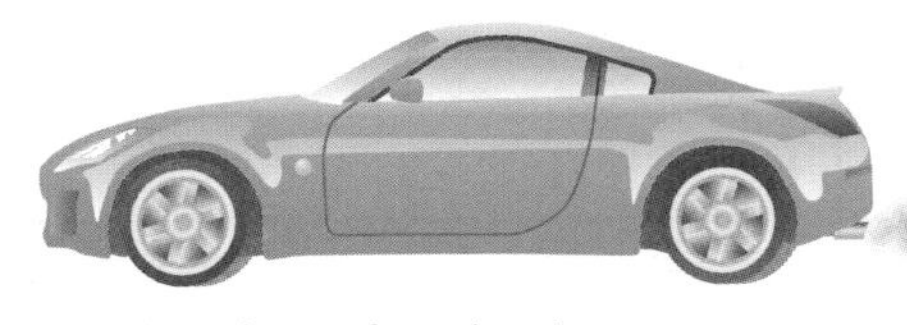

Out

- The normal products of combustion.
- **Carbon monoxide** – poisonous gas made when the fuel does not burn completely.
- **Oxides of nitrogen** – formed inside the internal combustion engine.

The table below includes information on how pollutant gases are produced, and what environmental problems they lead to. In your exam you may be asked to use information such as this to make judgements about the effects of air pollution, e.g. how it affects people's health.

Source	Gas Produced
Car exhausts	• Carbon monoxide (poisonous). • Nitrogen dioxide (leads to acid rain). • Unburned hydrocarbons (creates smog).
Oxides of nitrogen	• Acid rain and photochemical smog.
Aerosols	• CFCs (damage the ozone layer).

Burning fossil fuels (which are all carbon compounds) increases the amount of carbon dioxide in the atmosphere. However, this carbon dioxide can be used for photosynthesis.

It is important to reduce air pollution as much as possible because it can damage our surroundings and can adversely affect people's health. One way to remove carbon monoxide from car exhausts is to fit a **catalytic converter**. The catalyst causes the carbon monoxide to react, producing carbon dioxide.

Acid Rain

When coal or oils are burned, the sulfur impurities produce sulfur dioxide. Sulfur dioxide (and nitrogen dioxide) dissolves in water to produce acid rain. Acid rain can:

- erode stonework and corrode metals
- make rivers and lakes acidic and kill aquatic life
- kill plants.

HT Human Influence on the Atmosphere

Until relatively recently, the balance between adding and removing carbon dioxide from the atmosphere had remained constant. The levels of carbon dioxide and oxygen were maintained by photosynthesis and respiration. However, three important factors have upset the balance:

1. Excessive burning of fossil fuels is increasing the amount of carbon dioxide in the atmosphere.
2. Deforestation on large areas of the Earth's surface means the amount of photosynthesis is reduced so less carbon dioxide is removed from the atmosphere.
3. The increase in world population has directly and indirectly contributed to factors 1 and 2.

Oxygen and nitrogen from the air react together in the high temperature of a car engine to form oxides of nitrogen.

To help reduce the amount of pollutants being put into the atmosphere, catalytic converters are fitted to cars to convert the carbon monoxide in exhaust gases to the less harmful carbon dioxide.

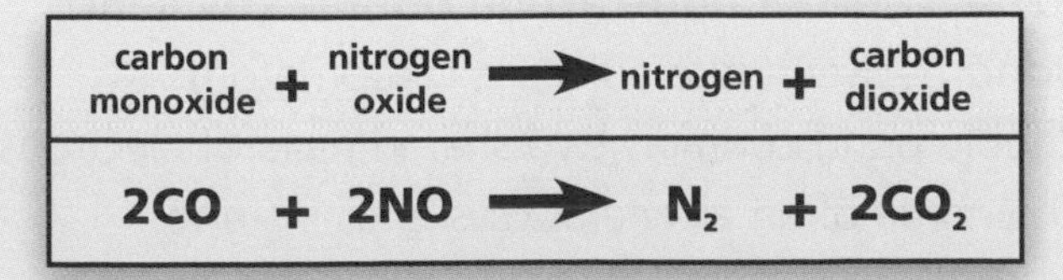

$$2CO + 2NO \rightarrow N_2 + 2CO_2$$

Hydrocarbons

Hydrocarbons are molecules that contain hydrogen and carbon atoms only.

You need to remember that:

- carbon atoms can make four bonds each
- hydrogen atoms can make one bond each.

Alkanes

When a hydrocarbon contains **single covalent bonds** only, it is called an **alkane**. The name of an alkane always ends in **-ane**.

This table shows the displayed and molecular formulae for the first four members of the alkane series.

Alkane	Displayed Formula	Molecular Formula
Methane	H | H – C – H | H	CH_4
Ethane	H H | | H – C – C – H | | H H	C_2H_6
Propane	H H H |	| H – C – C – C – H |	| H H H	C_3H_8
Butane	H H H H |		| H – C – C – C – C – H |		| H H H H	C_4H_{10}

Alkenes

The **alkenes** are another form of hydrocarbon. They are very similar to the alkanes except that they contain **one carbon–carbon double covalent bond** between two adjacent carbon atoms. A double bond contains two shared pairs of electrons. The name of an alkene always ends in **-ene**.

This table shows the displayed and molecular formulae for the first three members of the alkene series.

Alkene	Displayed Formula	Molecular Formula
Ethene	H H \ / C = C / \ H H	C_2H_4
Propene	H H \	 C = C – C – H /		 H H H	C_3H_6
Butene	H H H \	| C = C – C – C – H /		| H H H H	C_4H_8

A simple test to distinguish between alkenes and alkanes is to add bromine water. Alkenes react with bromine water (orange) and decolourise it (colourless). Alkanes have no effect on bromine water.

HT Alkanes contain only single covalent bonds between the carbon atoms – they are described as **saturated** hydrocarbons. (They have the maximum number of hydrogen atoms per carbon atom in the molecule.)

Alkenes contain at least one carbon–carbon double covalent bond. This means that the carbon atom is not bonded to the maximum number of hydrogen atoms. Alkenes are therefore described as being **unsaturated**.

Bromine water (orange) turns colourless when shaken with an alkene. This is an **addition reaction** as the bromine adds on to the alkene molecule to make a **colourless dibromo compound**, e.g.

$$C_2H_4 + Br_2 \longrightarrow C_2H_4Br_2 \text{ (1,2-dibromoethane)}$$

Polymerisation

The alkenes made by cracking can be used as **monomers**. Monomers are small molecules that can be reacted together to produce **polymers**. These are very large molecules, some of which make up plastics.

Alkenes are very good at joining together, and when they do so without producing another substance we call it **polymerisation**.

This process, e.g. the formation of poly(ethene) from ethene, requires **high pressure** and a **catalyst**.

The name of a polymer is made from the name of its monomer, e.g. ethene makes poly(ethene) and you can work out what the monomer is from the name of the polymer, e.g. poly(propene) is made from propene.

The small alkene molecules are called monomers.

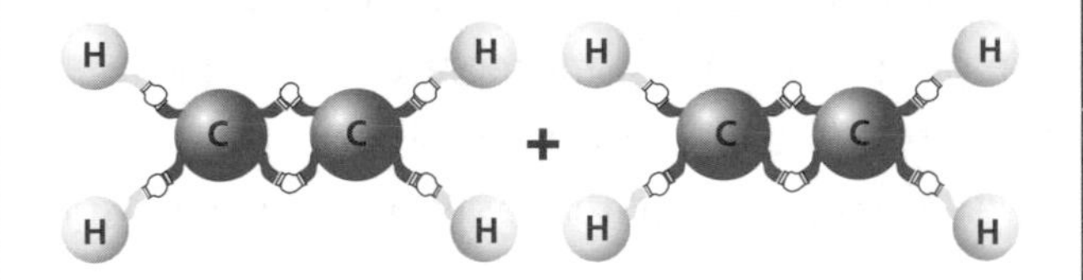

Their carbon–carbon double bonds are easily broken.

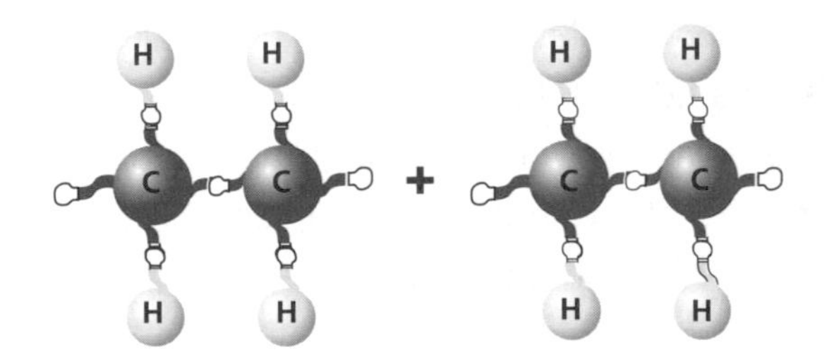

Large numbers of molecules can therefore be joined in this way. The resulting long-chain molecule is a polymer: poly(ethene) or polythene.

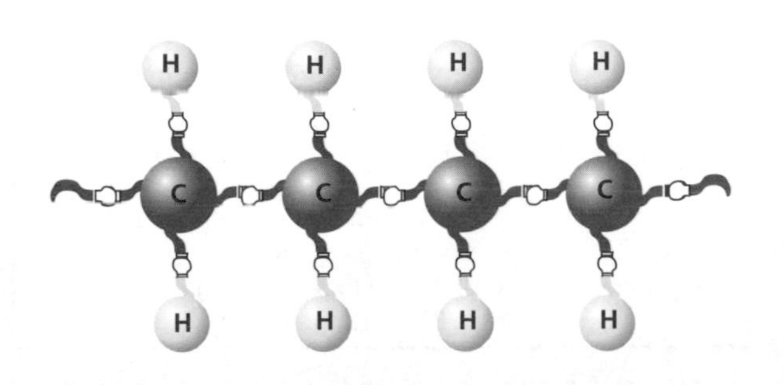

Consider the displayed formula of ethene and poly(ethene):

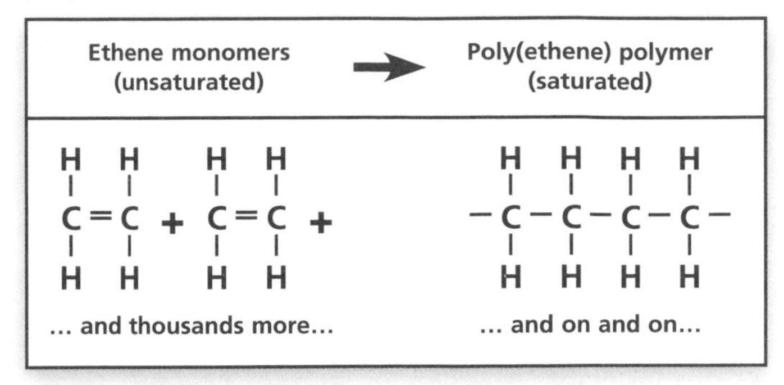

A more compact way of writing this reaction uses the standard way of displaying a polymer formula:

$$n\left(\begin{array}{c} H \quad\quad H \\ C=C \\ H \quad\quad H \end{array}\right) \rightarrow \left(\begin{array}{c} H \quad H \\ -C-C- \\ H \quad H \end{array}\right)_n$$

Addition polymerisation involves the reaction of many unsaturated monomer molecules, i.e. alkenes, to form a saturated polymer. You will be expected to be able to construct the displayed formula of:

- a **polymer** given the displayed formula of a monomer, e.g. propene monomer to poly(propene) polymer:

$$\begin{array}{c} H \quad\quad CH_3 \\ C=C \\ H \quad\quad H \end{array} \rightarrow \left(\begin{array}{c} H \quad CH_3 \\ -C-C- \\ H \quad H \end{array}\right)$$

- a **monomer**, given the displayed formula of a polymer, e.g. poly(propene) polymer to propene monomer:

$$\left(\begin{array}{c} H \quad CH_3 \\ -C-C- \\ H \quad H \end{array}\right) \rightarrow \begin{array}{c} H \quad\quad CH_3 \\ C=C \\ H \quad\quad H \end{array}$$

Designer Polymers

Polymers

Polymers (plastics) have many properties that make them useful. You should be able to use these terms to explain why a certain plastic is used for a particular job. Properties of plastics are listed below:

- Can be easily moulded into shape
- Waterproof
- Electrical insulator
- Non-biodegradable
- Lightweight
- Flexible
- Can be printed on
- Unreactive
- Can be coloured
- Heat insulator
- Transparent
- Tough

Uses for Polymers

Different plastics have different properties, which results in them having different uses:

Polymer	Properties	Uses
Polythene or poly(ethene)	• Lightweight • Flexible • Easily moulded	• Plastic bags • Moulded containers
Polystyrene (expanded polystyrene)	• Lightweight • Poor conductor of heat	• Insulation • Damage protection in packaging
Nylon	• Lightweight • Waterproof • Tough	• Clothing • Climbing ropes
Polyester	• Lightweight • Waterproof • Tough	• Clothing • Bottles

Outdoor Clothing

Outdoor clothing, such as a jacket, needs to be waterproof to keep the wearer dry. Nylon is an excellent material to use to make outdoor clothing because it is:

- lightweight
- tough
- waterproof (but it does not let water vapour escape, so it could be uncomfortable to wear if the wearer becomes hot and starts to perspire)
- able to block ultraviolet (UV) light (harmful sunlight).

Gore-Tex®

Gore-Tex® is a breathable material made from nylon. It has all of the advantages of nylon, but it is also treated with a material that allows perspiration (water vapour) to escape whilst preventing rain from getting in. This is far more comfortable for people who lead an active outdoor life, as it prevents them from getting wet when they perspire.

HT Gore-Tex® has a membrane of polyurethane or poly(tetrafluoroethane) (PTFE), sandwiched between two layers of nylon fibres. The laminated PTFE has very tiny holes that allows water vapour to pass through but that are too small to allow liquid water to pass through. The PTFE laminate is too weak and fragile to be used on its own.

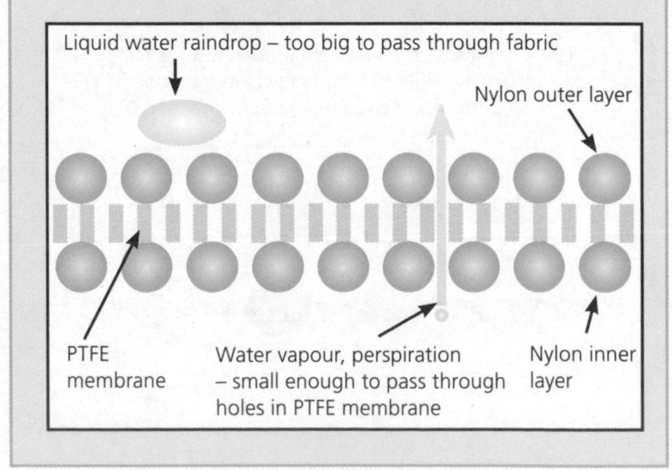

Structure of Plastics

Polymers (plastics), such as poly(chloroethene) (PVC), consist of a tangled mass of very long-chain molecules, in which the atoms are held together by strong covalent bonds. The properties of a plastic depend on its structure.

Plastics that have weak forces between polymer molecules have low melting points and can be stretched easily as the polymer molecules can slide over one another.

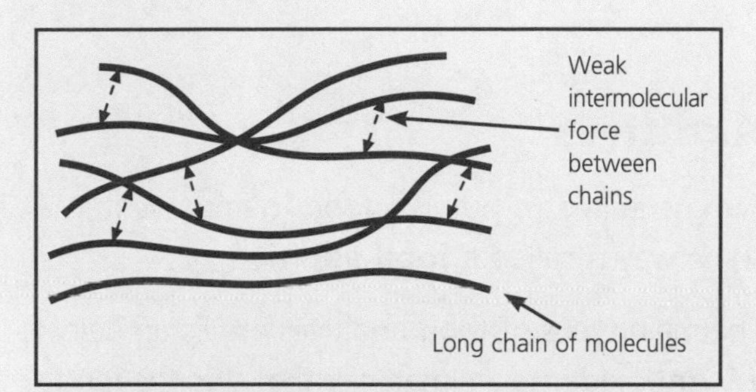

Plastics that have strong forces between the polymer molecules (covalent bonds or cross-linking bridges) have high melting points, are rigid and cannot be stretched.

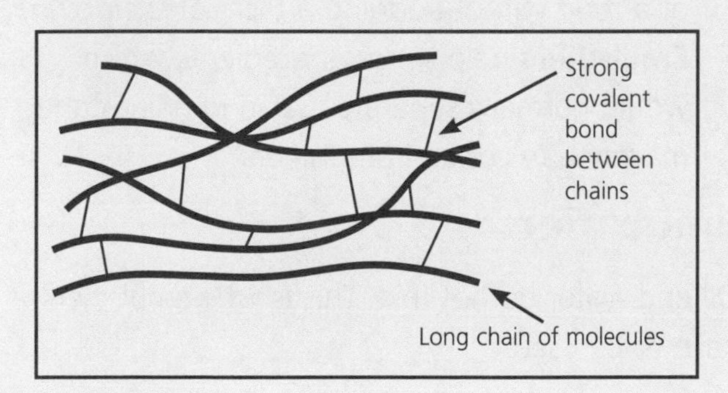

Disposal of Plastics

As we have seen, plastics have many different uses. As it is such a convenient material, we produce a large amount of plastic waste. This can be difficult to dispose of and can sometimes be seen as litter in the streets. There are various ways of disposing of plastics. Unfortunately, some of them have a negative impact on the environment:

- Using **landfill sites** is a problem because most plastics are non-biodegradable. This means microorganisms have no effect on them and they will not decompose and rot away. Throwing plastics into landfill sites results in the waste of a valuable resource and, because of the volume of waste produced, landfill sites get filled up very quickly, which is also a waste of land.
- **Burning** plastics produces air pollution and also wastes valuable resources. The production of carbon dioxide contributes to the greenhouse effect which results in global warming. Some plastics cannot be burned at all because they produce toxic fumes. For example, burning poly(chloroethene) or PVC as it is more commonly known, produces hydrogen chloride gas.
- **Recycling** plastics is an option which prevents resources being wasted. However, different types of plastic need to be recycled separately. Sorting them into groups to be recycled can be difficult and very time-consuming.

Research is being carried out on the development of biodegradable plastics to help reduce the impact that the disposal of plastics has on the environment.

Soluble plastics make disposal easy. For example, the plastic case of a dishwasher tablet is disposed of when it dissolves in contact with hot water, releasing detergent into the machine.

C1 Cooking and Food Additives

Cooking Food

Cooking food causes a **chemical change** to take place. When a chemical change occurs:

- new substances are formed from old ones
- there may be a change in mass when a gas is released
- there is often a substantial energy change, e.g. a rise or fall in temperature
- the change cannot be reversed easily.

Cooking Eggs and Meat

Eggs and meat contain lots of protein. The protein molecules change shape when they are heated. This is called **denaturing**.

HT The texture of eggs and meat changes when they are cooked because the protein molecules change shape permanently.

Potatoes and Vegetables

Potatoes and other vegetables are plants; their cells have a rigid cell wall. During cooking, the heat breaks down this cell wall, starch is released and it becomes much softer. The starch grains swell up and spread out, and the potato is now much easier to digest.

Baking Powder

Baking powder contains sodium hydrogencarbonate. When this is heated, it breaks down (decomposes) to make sodium carbonate and water, and carbon dioxide gas is given off.

The word equation for the decomposition of sodium hydrogencarbonate is:

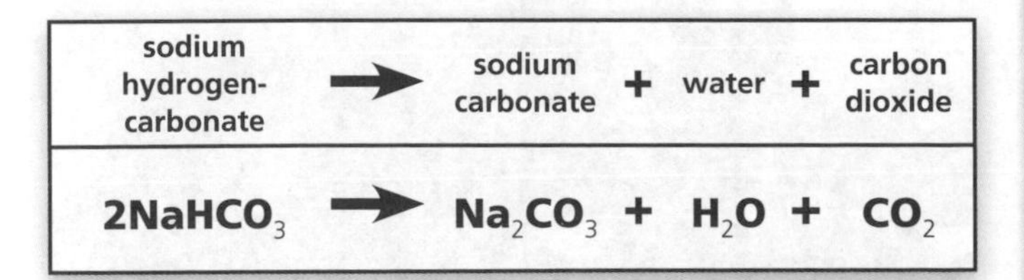

sodium hydrogen-carbonate	→	sodium carbonate	+	water	+	carbon dioxide
$2NaHCO_3$	→	Na_2CO_3	+	H_2O	+	CO_2

Baking powder is added to cake mixture because the carbon dioxide gas given off when it is heated causes the cake to **rise**.

You can test for the presence of carbon dioxide using **limewater** (**calcium hydroxide solution**). If carbon dioxide is present, the limewater turns **milky**.

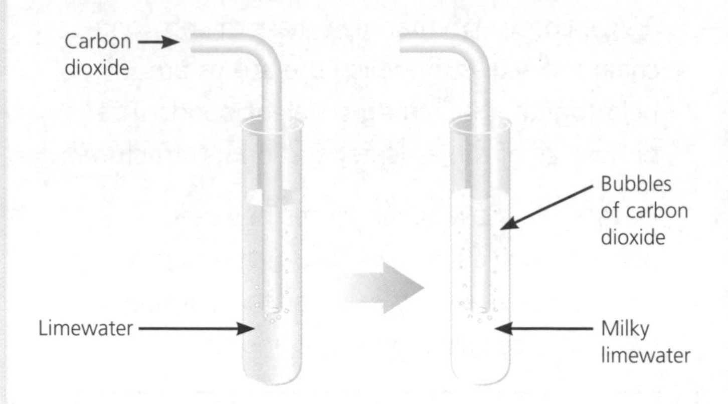

Additives

A material that is put in a food to improve it in some way is called a **food additive**.

The main types of food additives are listed below:

- **Antioxidants** are materials that stop the food reacting with oxygen in the air. They are usually added to foods that contain fats or oils, e.g. bacon.
- **Food colours** improve the appearance of food.
- **Flavour enhancers** help bring out the flavour of a food without adding a taste of their own.
- **Emulsifiers** help to mix ingredients which would normally separate. Salad dressings and mayonnaise contain emulsifiers.

Emulsifiers

Oil and water do not mix. This is why emulsifiers have to be used.

The molecules in an emulsifier have two ends: one end likes to be in water (**hydrophilic**) and the other end likes to be in oil (**hydrophobic**). The emulsifier joins the droplets together and keeps them mixed.

HT The **hydrophilic** end of the emulsifier molecule bonds to the **polar water molecules**. The **hydrophobic** end of the emulsifier molecule bonds to the **non-polar oil molecules**.

Perfumes

Smells are made of molecules which travel up your nose and stimulate sense cells.

A perfume must smell nice. In addition, it must:

- evaporate easily – so it can travel to your nose
- not be toxic – so it does not poison you
- not irritate – otherwise it would be uncomfortable on the skin
- not dissolve in water – otherwise it would wash off easily
- not react with water – otherwise it would react with perspiration.

There are many kinds of perfume. Some come from **natural** sources, such as plants and animals. Perfumes can also be **manufactured**. If they are manufactured they are known as **synthetic** perfumes.

Esters are a common family of compounds used as synthetic perfumes. An ester is made by reacting an alcohol with an organic acid. This produces an ester and water. A simple ester, ethyl ethanoate, is made by adding ethanoic acid to ethanol (see below).

Perfumes and cosmetics need to be tested to make sure they are not harmful. This testing is sometimes done on animals. Some people are not happy about this. They argue that it is cruel to animals, and pointless because animals do not have the same body chemistry as humans and so results of the tests may not be useful. However, the tests could be useful to prevent humans from being harmed. The testing of cosmetics on animals is now banned in the EU.

HT Perfumes are **volatile**: they evaporate easily.

The molecules in a drop of perfume are held together by weak intermolecular forces of attraction. The molecules that escape have lots of energy and easily overcome the weak attraction to the other molecules in the liquid.

Making a Simple Ester

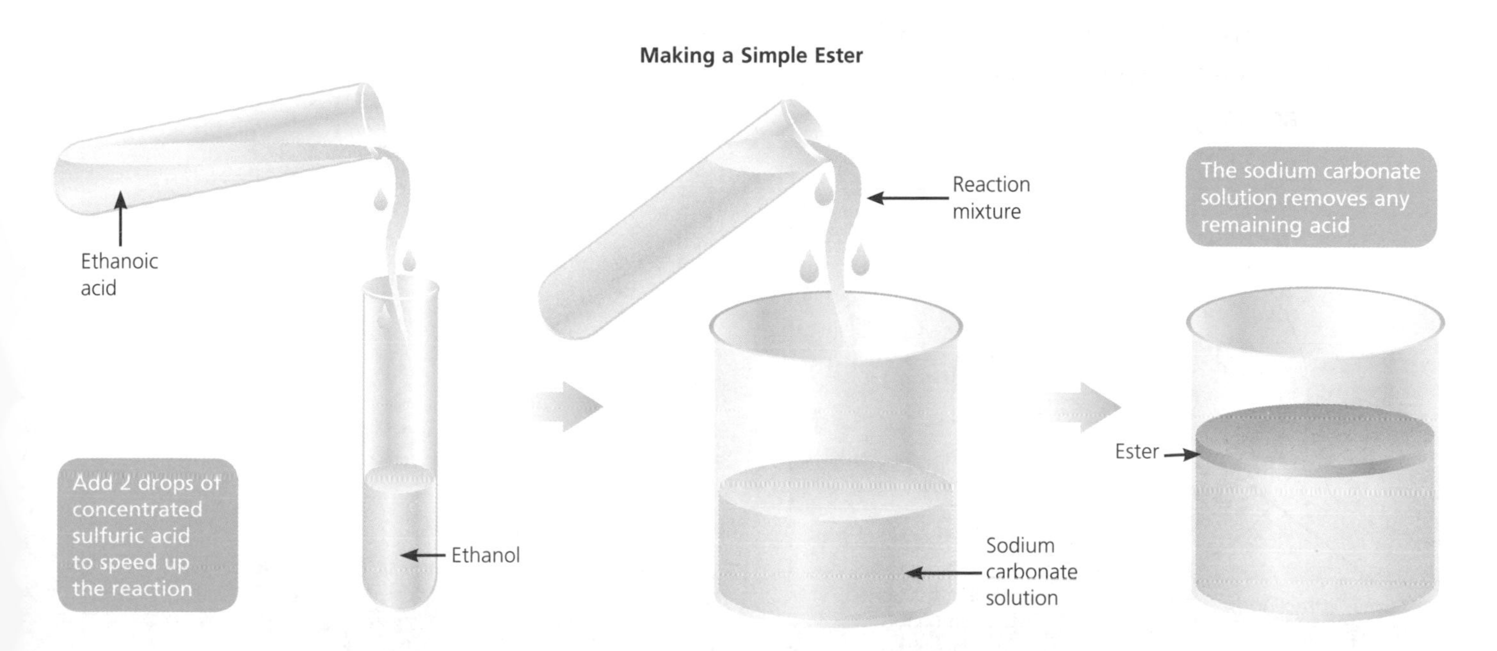

Solvents

Below are some words that are used to describe substances (together with their definitions):

- **Soluble substances** are substances that dissolve in a liquid, e.g. nail varnish is soluble in ethyl ethanoate.
- **Insoluble substances** are substances that do not dissolve in a liquid, e.g. nail varnish is insoluble in water.
- A **solvent** is the liquid into which a substance is dissolved, e.g. ethyl ethanoate is a solvent. (An ester can be used as a solvent.)
- The **solute** is the substance that gets dissolved, e.g. the nail varnish is a solute.
- A **solution** is what you get when you mix a solvent and a solute. The mixture does not separate out.

Nail varnish colours dissolve in nail varnish remover (a solvent).

HT Water will not dissolve nail varnish because:

- the attraction between water molecules is stronger than the attraction between water molecules and the particles in nail varnish
- the attraction between the particles in nail varnish is stronger than the attraction between water molecules and particles in nail varnish.

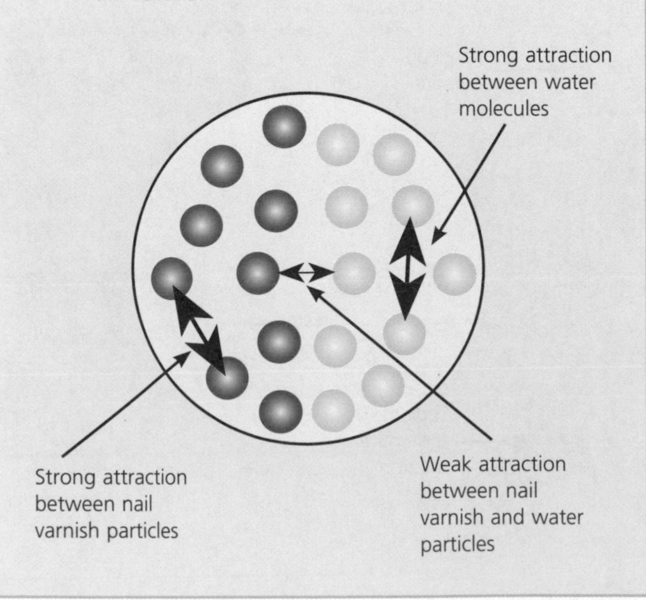

You may be asked to put solvents in order of most effective to least effective in removing nail varnish, or to suggest which solvent you could use to dissolve magnesium chloride (a salt).

You may be asked to decide how good a solvent is from information collected in an experiment, such as in this table:

Solvent	Nail Varnish	Sodium Chloride (Salt)
Water	Does not dissolve	Very soluble
Ethanol	Dissolves in 15 seconds	Slightly soluble
Ethyl ethanoate	Dissolves in 3 seconds	Insoluble
Propanone	Dissolves in 2 seconds	Insoluble

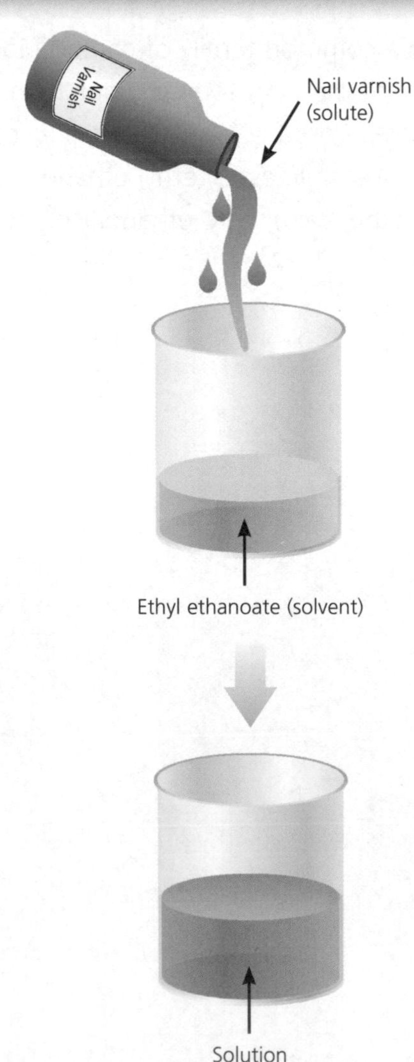

Paint

Paint can be used to protect or to decorate a surface. It is a special mixture of different materials, and it is called a **colloid**. In a colloid, fine solid particles are well mixed (dispersed) with liquid particles but they are not dissolved.

Paint is a mixture of:

- a **pigment** – a substance that gives the paint its colour
- a **binding medium** – an oil that sticks the pigment to the surface it is being painted onto
- a **solvent** – dissolves the thick binding medium and makes it thinner and easier to coat the surface.

The paint coats a surface with a thin layer and the solvent evaporates away as the paint dries. The solvent in **emulsion** paint is **water**. In **oil-based** paints, the pigment is dispersed in an **oil** (the binding medium). Often, there is a solvent present that dissolves the oil.

Special Pigments

Thermochromic pigments change colour when they are heated or cooled. These pigments can be used in:

- cups and kettles to warn that they are hot
- mood rings that change colour as your body temperature changes
- baby feeding spoons to show that the food is not too hot
- bath toys to show that the water is the correct temperature for a baby.

Phosphorescent pigments glow in the dark. They absorb and store energy and release it slowly as light when it is dark. The paint on some watch dials contains phosphorescent pigments.

HT More about Paint

The particle size of the solids in a colloid must be very small so they stay scattered throughout the mixture. If the particles were too big, they would start to settle down to the bottom.

An oil-based paint such as a gloss paint dries in two stages:

1. The solvent **evaporates** away.
2. The oil-binding medium reacts with oxygen in the air as it dries to form a hard layer. This is an **oxidation** reaction.

More about Pigments

There are only a few thermochromic pigments. To increase the range of colours, they can be mixed with ordinary pigments in acrylic paints.

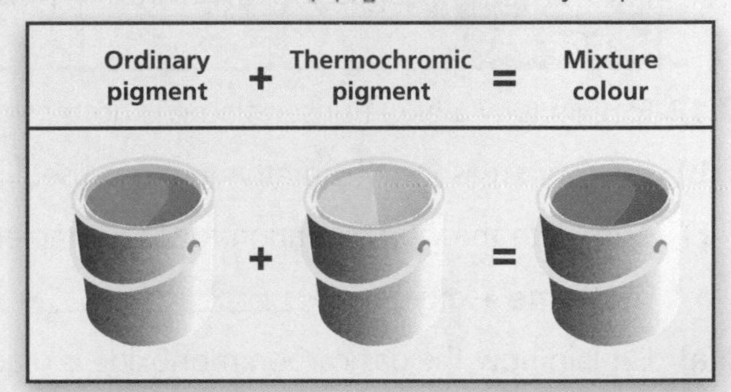

Thermochromic pigments change to colourless as they get hotter and so the paint changes from the mixture colour to the ordinary pigment colour.

The first 'glow in the dark' paints were made using radioactive materials as pigments. They were used to paint the dials on aircraft instrument panels and the first luminous watches. However, the people who painted with these pigments were exposed to too much radiation and some of them developed cancer as a result. Phosphorescent pigments are not radioactive, so they are much safer to use.

Exam Practice Questions

1 a) Name the three main fossil fuels. [3]

b) Explain why diesel fuel is non-renewable. [2]

c) An oil company wants to build an oil tanker terminal near a large sea bird colony. Write about the advantages and disadvantages of building a terminal in this area. [6]

The quality of your written communication will be assessed in your answer to this question.

d) How does fractional distillation separate the components of crude oil? [2]

e) Describe how cracking helps to meet the demand for petrol. [3]

f) Look at the formulae of these molecules:

A CH_4 **B** C_3H_6 **C** C_4H_{10} **D** C_2H_5OH **E** C_3H_8

i) Which molecule is not a hydrocarbon? [1]

ii) Which molecule contains 9 atoms? [1]

g) Benzene is found in crude oil. It has a boiling point of 81°C. After distillation, which of the fractions of crude oil will contain benzene? Use the information in the table to help you. [1]

Fraction	Refinery gases	Petrol	Paraffin/ heating oil	Diesel	Lubricating oil	Fuel oil	Bitumen
Boiling range (°C)	Up to 25	40–100	150–250	220–350	Over 350	Over 400	Over 400

2 a) Explain why a blue Bunsen flame is hotter than a yellow Bunsen flame. [2]

b) Give two reasons why methane gas is used in many homes as a fuel for cooking. [2]

c) Complete the word equation for the complete combustion of methane.

methane + oxygen → ____________ + ____________ [1]

3 a) Explain how the gas carbon monoxide is made in a car engine and why it is air pollution. [2]

b) Suggest how photosynthesis changes the amounts of oxygen and carbon dioxide in the air. [2]

c) Describe **three** effects of acid rain. [3]

4 a) Explain what 'non-biodegradeable' means. [1]

b) What are the two main properties that make Gore-Tex® so useful as a material for jackets for hill walkers? [2]

5 a) An ester is made by this reaction:

ethanol + ethanoic acid → ethyl ethanoate + water

Give the name of a reactant in this reaction. [1]

b) A scientist has developed a new paint mixture and she now wants to try different colours. What should she change to get a range of different colours? [1]

HT

6 Explain why small molecule hydrocarbons have a lower boiling point than larger molecule hydrocarbons. [2]

7 Write the formula equation for the complete combustion of methane. Include state symbols. [3]

8 Explain how deforestation affects the Earth's atmosphere. [5]

9 a) Write the formula equation for the thermal decomposition of baking powder. [2]

b) What happens to a protein molecule when it is denatured? [1]

C2: Chemical Resources

This module looks at:

- The structure of the Earth, volcanoes and the theory of plate tectonics.
- How rocks are used as the raw materials for the construction industry.
- Metal properties and alloys, and how metals are extracted from ores.
- The materials used in car making, and comparisons between aluminium and iron.
- The manufacture of ammonia and the conditions used in the Haber process.
- Acids, bases and neutralisation reactions.
- The benefits and disadvantages of using chemical fertilisers.
- The electrolysis of sodium chloride as a source of important materials.

Structure of the Earth

The Earth is nearly spherical and has a layered structure as shown below.

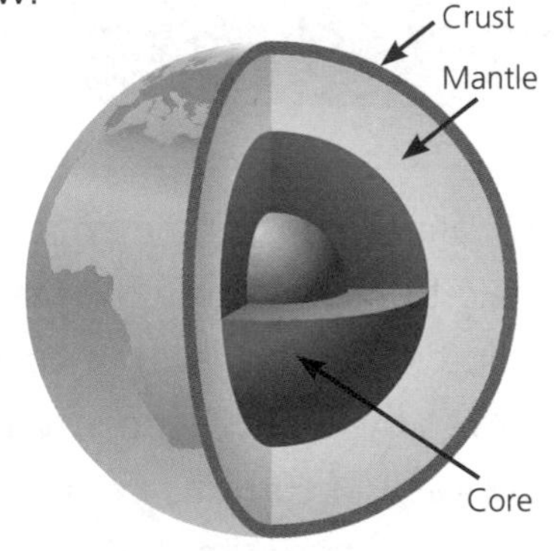

The thickness of the thin, rocky **crust** varies between 10km and 100km. **Oceanic** crust lies beneath the oceans. **Continental** crust forms the continents.

The **mantle** extends almost halfway to the centre of the Earth. It has a higher density than rock in the crust, and has a different composition.

The **core** accounts for over half of the Earth's radius. It is mostly made of iron.

It is difficult to collect information about the structure of the Earth. The deepest mines and deepest holes drilled into the crust have penetrated only a few kilometres. Scientists have to rely on studying the seismic waves (vibrations) caused by earthquakes and man-made explosions.

Movement of the Lithosphere

The Earth's **lithosphere** is the relatively cold, rigid, outer part of the Earth, consisting of the crust and outer part of the mantle. The top of the lithosphere is 'cracked' into several large interlocking pieces called **tectonic plates.**

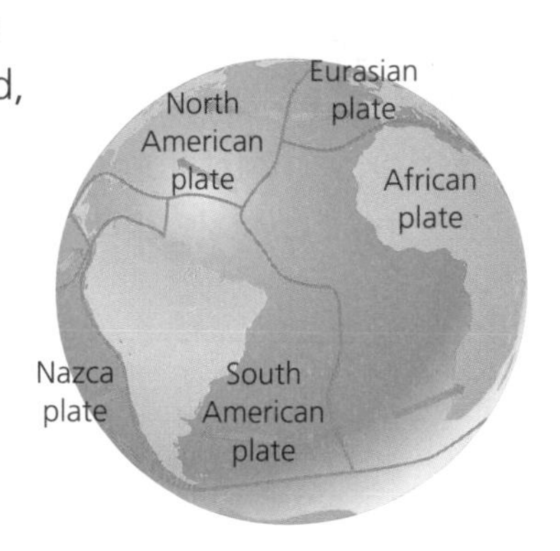

Many ideas have been used to explain how the Earth's surface behaves, but most scientists now accept the theory of plate tectonics. It fits with a wide range of evidence and many scientists have discussed and tested it.

The plates sit on top of the mantle because they are less dense than the mantle itself. Although there does not appear to be much going on, the Earth and its crust are very dynamic. They move very slowly, at speeds of about 2.5 cm a year. Plates can move apart from, towards, or slide past, each other. This movement causes **earthquakes** and **volcanoes** at the boundaries between plates. It has taken millions of years for the continents to have moved to where they are today.

Volcanoes

Volcanoes form where molten rock can find its way through to the Earth's surface, usually at plate boundaries or where the crust is weak.

Volcanoes can give out lava that can move fast and is very runny, or erupt thick lava violently with disastrous effects. Living near a volcano can be very dangerous, but people often choose to live there because volcanic soil is very fertile.

Geologists study volcanoes to help understand the structure of the Earth and also to help predict when eruptions will occur to give an early warning for people who live nearby. However, they still cannot predict with 100% certainty.

Volcanoes (cont)

Igneous rock is formed when molten rock cools down. Igneous rocks are hard and have interlocking crystals:

- Large crystals form when molten rock cools slowly.
- Small crystals form when molten rock cools quickly.

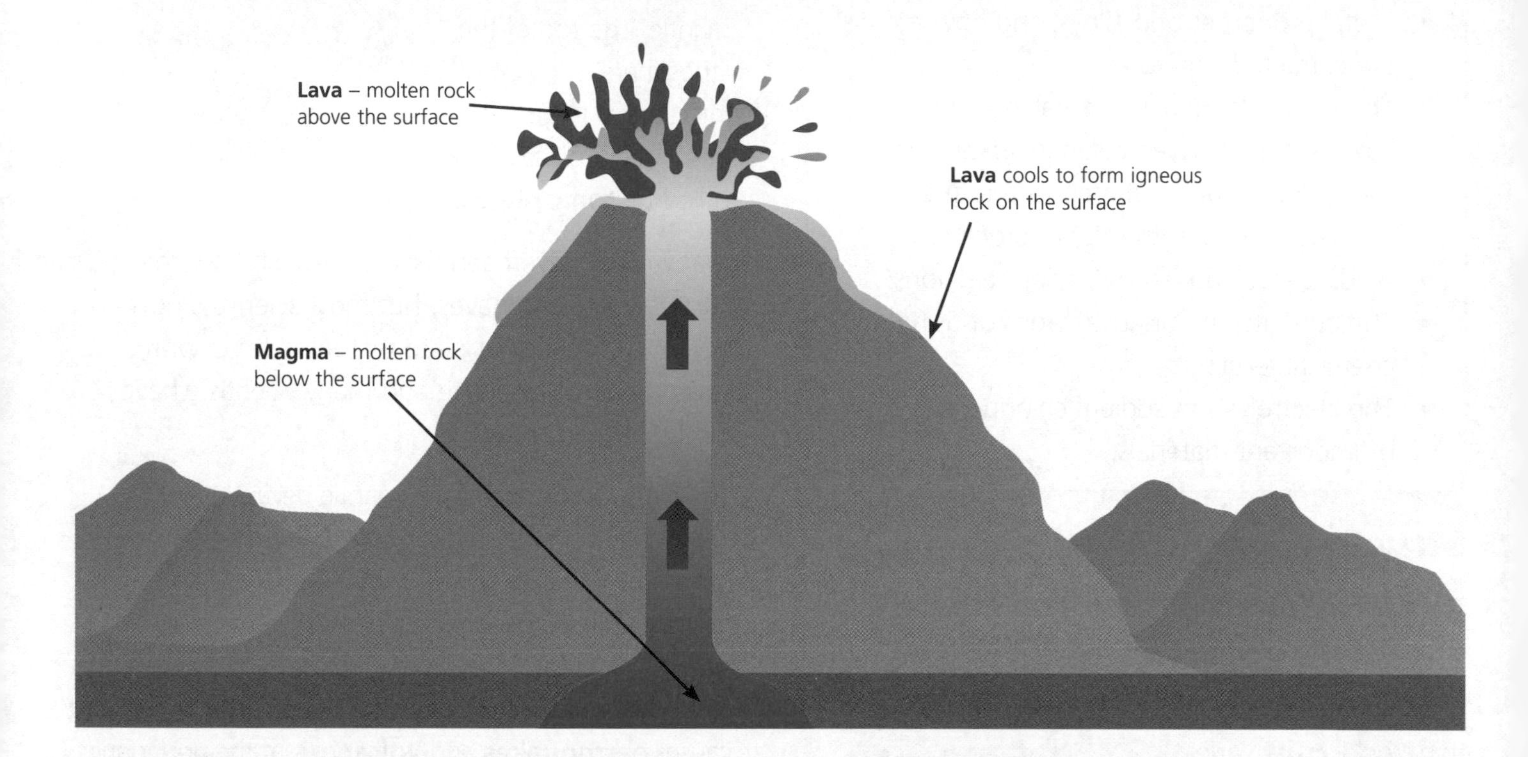

Magma

The different compositions of magma affect the type of rock that will form, and the type of eruption.

Iron-rich basalt magma is quite runny and fairly safe in comparison with silica-rich rhyolite, which is thicker, and treacle-like. The volcanoes that have thicker magma can erupt violently.

Geologists are getting better at forecasting volcanic eruptions but they still cannot predict with 100% certainty.

What Causes Plates to Move?

In the zone below the lithosphere and above the core, the mantle is relatively cold and rigid. At greater depths, the mantle is hot, non-rigid and able to flow. Within the mantle, at these greater depths, are **convection currents** which are driven by heat released from radioactivity. These convection currents in the semi-rigid mantle transfer energy to the plates which then move slowly.

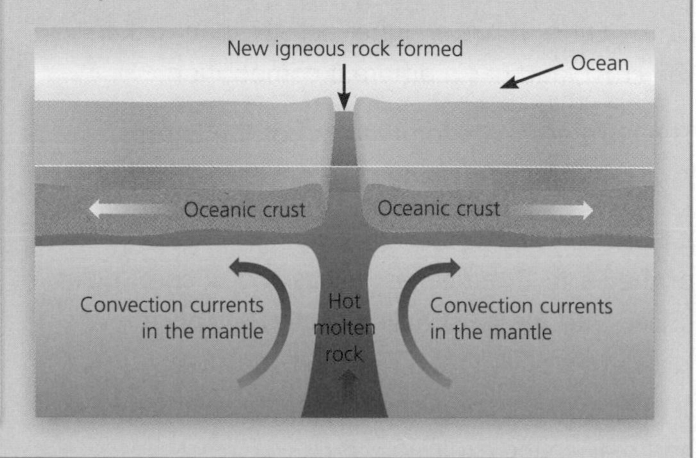

Effects of Plate Collision

While some plates around the world are moving **apart**, others are moving **towards** each other. An oceanic plate 'dips down' when it collides with a continental plate and slides under it, because oceanic crust has a higher density than continental crust. This is called **subduction**.

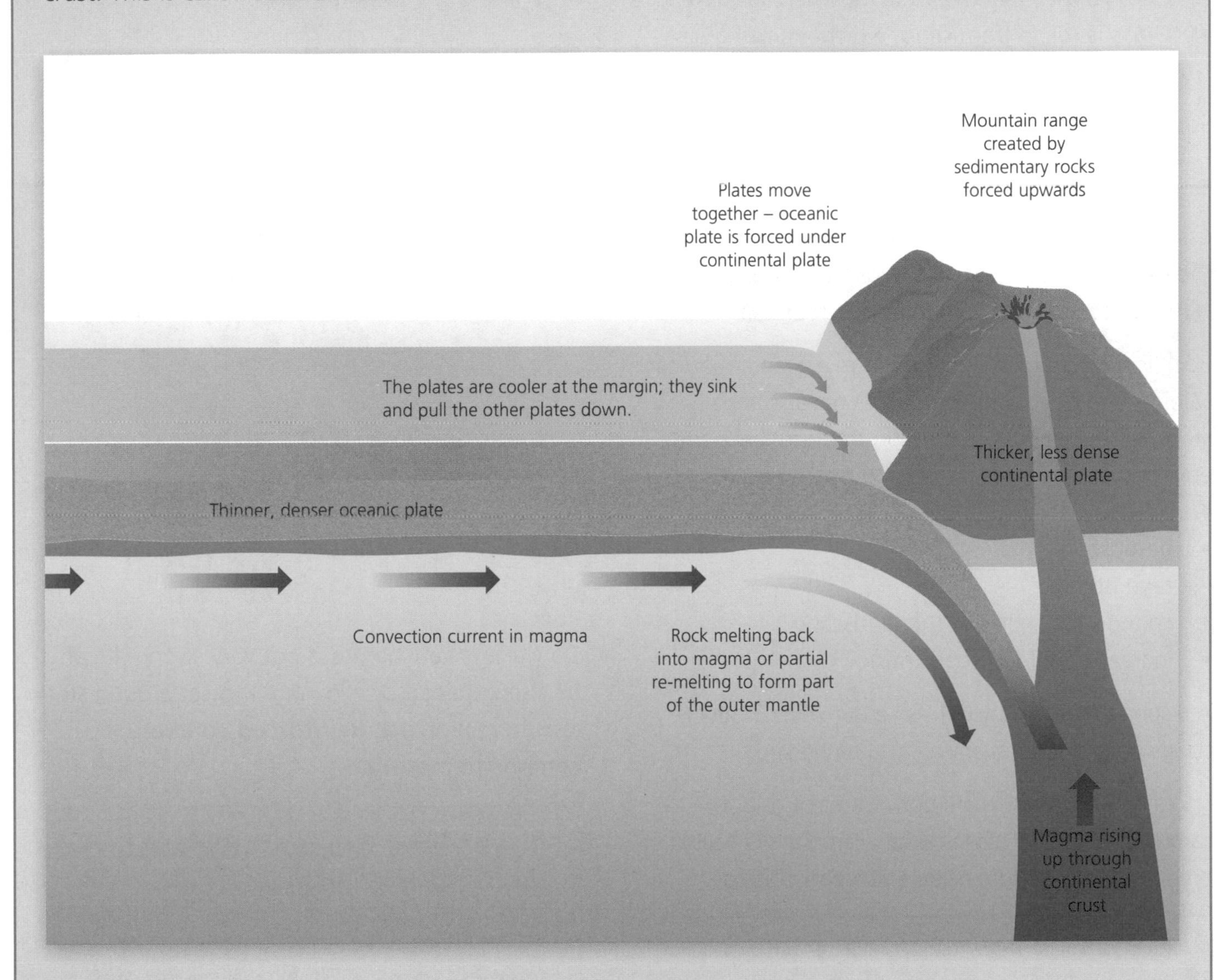

In 1914 Alfred Wegener put forward a theory that the continents were drifting apart. Some of the evidence he used included:

- The continents seemed to fit together like a jigsaw.
- Fossils in Africa matched those in South America.
- Greenland was getting further away from Europe.

Wegener's ideas were not accepted by the majority of scientists until new evidence came to light in the 1960s from the study of the formation of new rocks at the sea floor either side of where a crack occurred. This and subsequent research has backed up his theory, which developed into the theory of plate tectonics. The theory has now gained acceptance by the scientific community.

C2 Construction Materials

Materials from Rocks

Many construction materials come from substances found in the Earth's crust:

- The metals **iron** and **aluminium** are extracted from rocks called ores.
- Clay is a rock that makes **brick** when it is baked.
- **Glass** is made from sand, which is small grains of rock.

Some rocks, like limestone, marble and granite, just need to be shaped to be ready to use as a building material. **Aggregates** (crushed rock or gravel) are used in road making and in building. Limestone is the easiest to shape because it is the softest; marble is harder to shape and granite is harder still.

Rock is dug out of the ground in mines and quarries. Mining and quarrying companies have to take steps to reduce their impact on the local area and environment because mines and quarries can:

- be noisy
- be dusty
- take up land
- change the shape of the landscape
- increase the local road traffic.

A responsible company will also ensure it reconstructs, covers up and restores any area it has worked on.

Limestone and marble are both forms of calcium carbonate ($CaCO_3$). When calcium carbonate is heated it breaks up into calcium oxide and carbon dioxide.

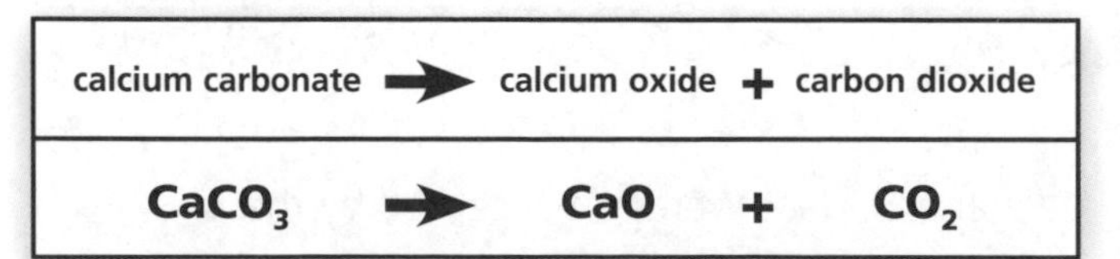

calcium carbonate	→	calcium oxide	+	carbon dioxide
$CaCO_3$	→	CaO	+	CO_2

This type of reaction is called a **thermal decomposition**; one material breaks down into two new substances when it is heated.

When clay and limestone are heated together, **cement** is made. One of the main uses of cement is to make **concrete**.

Concrete

To make concrete you need:

- 1 bag of cement
- 5 bags of sand
- 5 bags of aggregate
- water.

1. Mix all dry ingredients.
2. Add water.
3. Mix well.
4. Allow to set.

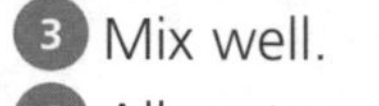

Concrete is very hard but not very strong. It can be strengthened by allowing it to set around steel rods to reinforce it. **Reinforced concrete** is a **composite material**.

HT Rocks and Composite Materials

Rocks differ in hardness because of the ways in which they were made. Limestone is a **sedimentary** rock. Marble is a **metamorphic** rock made from limestone that has been put under pressure and heated, which makes it harder. The hardest rock, granite, is an **igneous** rock.

A composite material combines the best properties of each material. Reinforced concrete combines the strength and flexibility of the steel bars with the hardness and bulk of the concrete. Reinforced concrete has many more uses than ordinary concrete.

Copper

Copper is a very useful metal because it is an excellent conductor of heat and electricity and it does not corrode. Copper is made by heating naturally occurring copper ore with carbon. The reaction removes oxygen from the ore; this is a **reduction** reaction.

The process uses lots of energy, which makes it expensive. It is cheaper to recycle copper than to extract it from its ore. Recycling also conserves our limited supply of copper ore and uses less energy. However, there are some problems with recycling copper. It has to be separated from other metals or there may be metals mixed with it and it is not always easy to persuade the public to recycle materials.

Electrolysis

New and recycled copper is purified by the process of **electrolysis**. The diagram below shows the apparatus used to purify copper:

HT **Electrolysis** is the name given to a chemical reaction that uses electricity. The electricity is passed through a liquid or a solution called an **electrolyte**, e.g. copper(II) sulfate solution is used to purify copper. **Electrodes** are used to connect to the electrolyte. The positive electrode (**anode**) is made of impure copper and the negative electrode (**cathode**) is made of pure copper.

The cathode increases in mass because pure copper is deposited on it from the electrolyte.

$$Cu^{2+} + 2e^- \longrightarrow Cu$$

This is a reduction reaction because the copper ion has gained electrons.

The anode loses mass as the copper dissolves into the electrolyte.

$$Cu - 2e^- \longrightarrow Cu^{2+}$$

This is an oxidation reaction because the copper atom has lost electrons.

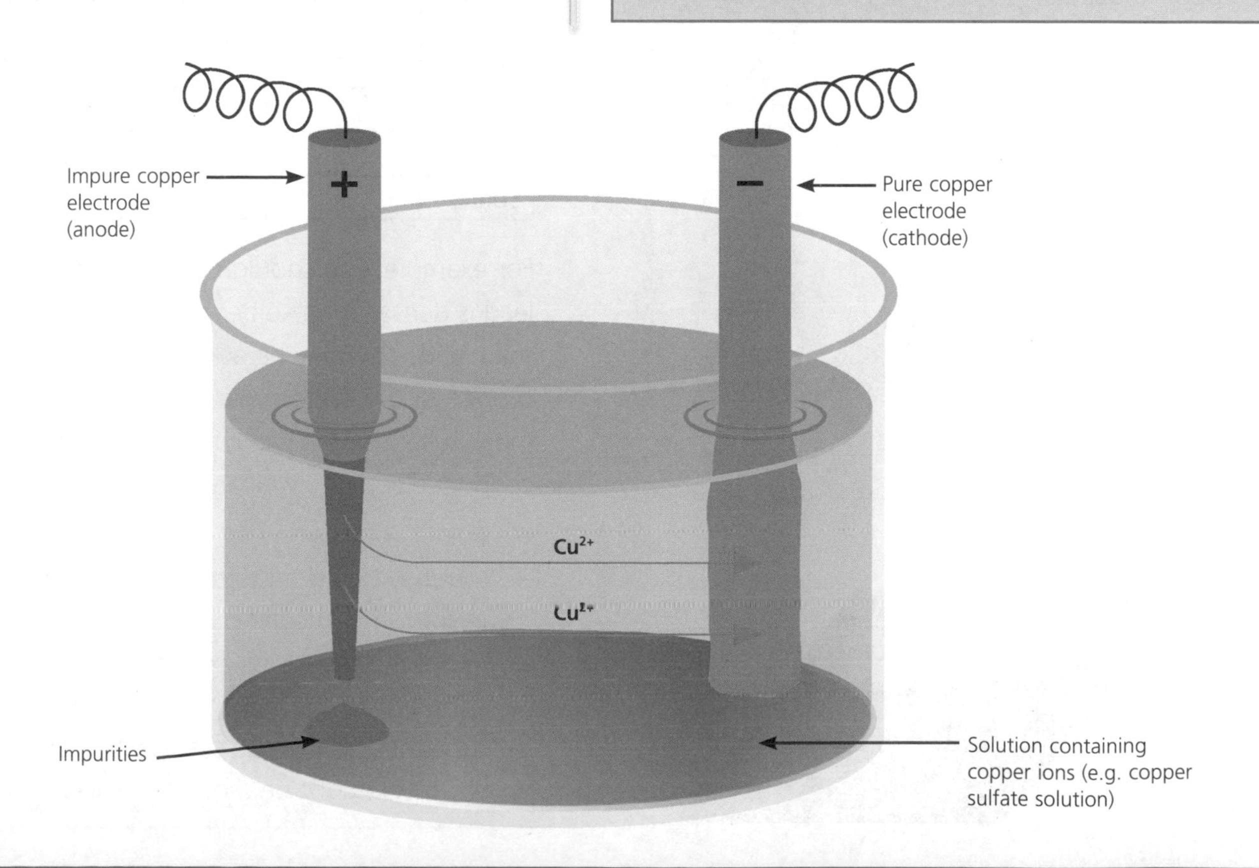

Alloys

An **alloy** is a mixture of a metal with another element (usually another metal). Alloys, e.g. bronze and steel, are made to improve the properties of a metal and to make them more useful – they are often harder and stronger than the pure metal:

- **Amalgam** is made using mercury and is used for fillings in teeth.

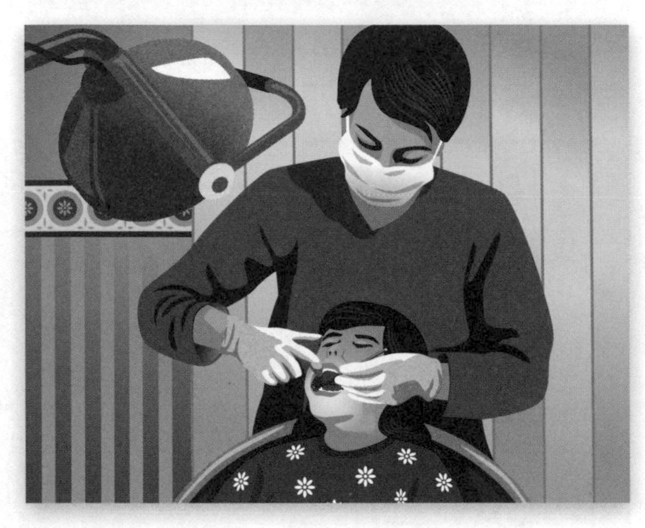

- **Brass** (made of copper and zinc) is used in coins, musical instruments, door handles and door knockers.

- **Solder** (made of lead and tin) is used to join electrical wires.

HT Smart Alloys

A **smart alloy** such as **nitinol** (an alloy of nickel and titanium) is used to make spectacle frames because it can be bent and twisted but it will return to its original shape when it is heated. It has **shape memory**.

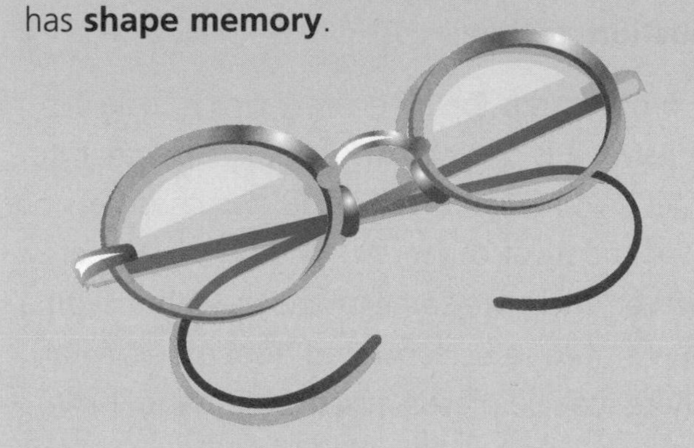

Metals and Properties

You may be asked to compare metals and alloys from a table such as this.

	Order of Hardness	Density (g/cm^3)	Melting Point (°C)	Order of Strength
Copper	4	8.9	1083	4
Brass	2	8.6	920	2
Steel	1	7.8	1420	1
Lead	5	11.3	327	5
Solder	3	9.6	170	3

For example, you should be able to work out that lead is denser than solder but that solder is stronger than lead. You may be asked to suggest which metal properties are important for a particular use, e.g. the metal wire inside a vacuum cleaner cable should conduct electricity well and be flexible.

HT You may be asked to explain why a metal is used for a particular purpose given data such as that in the table. For example, giving more than one reason why brass could be used to make the cables holding up a suspension bridge, or why steel might be a better choice.

Rusting Conditions

The diagram opposite shows an investigation into what conditions are needed to make a nail rust. Four nails were placed in test tubes in different conditions and left for a week. The only nail that rusted was the one in test tube 3. From this we can tell that rusting needs iron, water and oxygen (in air). The addition of oxygen to the iron is an **oxidation reaction**.

Rust flakes off the iron exposing more metal to corrode. Rusting happens even faster when the water is salty or is made from acid rain. Car bodies can rust. They are usually scrapped when this happens because it makes the metal weaker. Aluminium does not rust or corrode in air and water. Instead, it quickly forms a layer of aluminium oxide when it comes into contact with air. This layer stops any more air or water from coming into contact with the metal. This built-in protection will not flake off.

Rusting is an example of an oxidation reaction. This is a reaction where oxygen is added to a substance to make an oxide. Oxygen is added to the iron in the presence of water.

iron + oxygen + water ⟶ hydrated iron(III) oxide

Only iron and steel rust; other metals corrode. In your exam you may be asked to interpret information about the rate of corrosion of different metals. For example, the table below gives descriptions of metals in different conditions. You could state that aluminium is the least corroded metal because even in salty wet air, its appearance is the least changed (it has only dulled).

Rusting Investigation

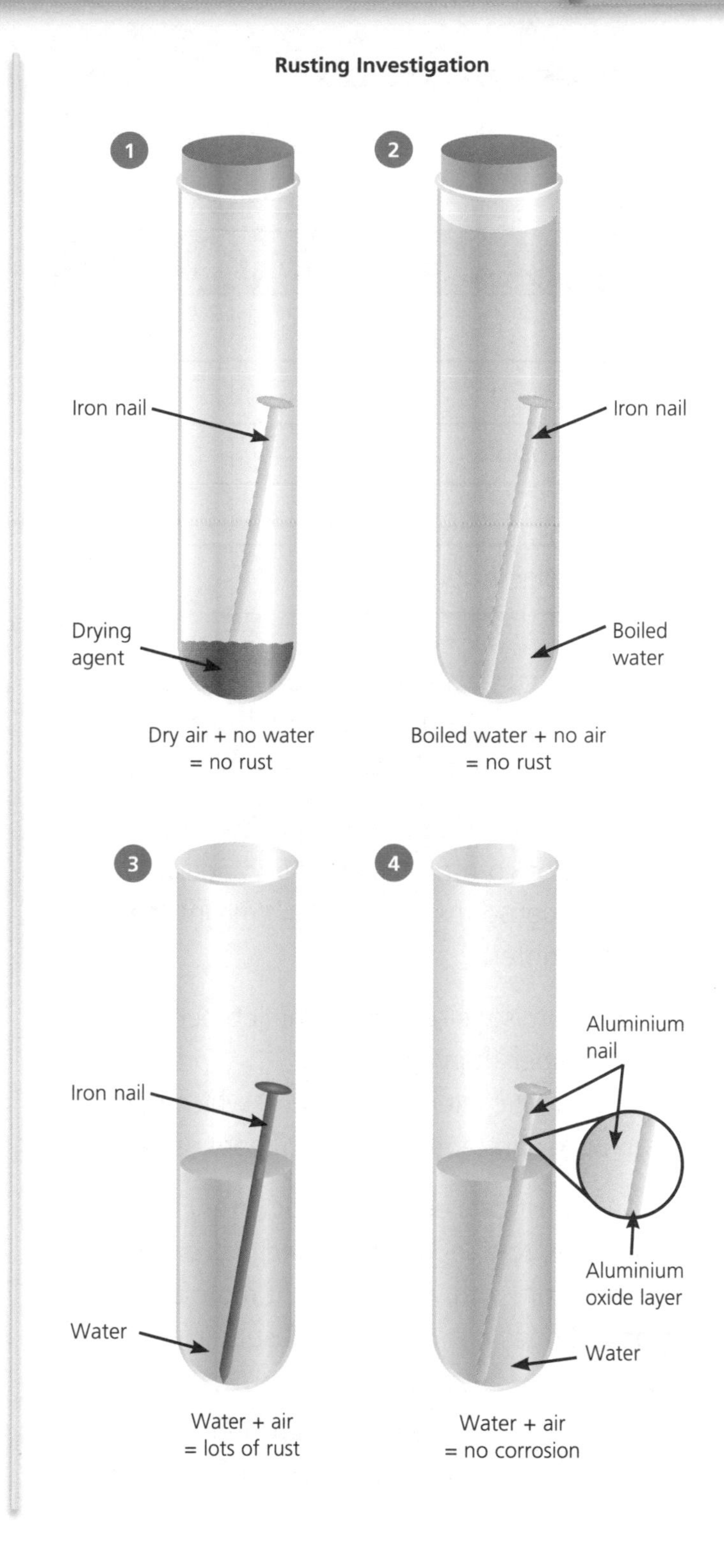

Metal	Clean and Dry Air	Wet Air	Acidic Wet Air	Salty Wet Air
Steel	Shiny	Dull (rusty)	Very dull (rusty)	Very dull (rusty)
Copper	Shiny	Dull	Green layer	Green layer
Aluminium	Shiny	Shiny	Dull	Dull
Silver	Shiny	Shiny	Tarnished	Tarnished

Properties of Metals

The table below shows the properties of aluminium and iron:

Property	Aluminium	Iron
Dense	✘	✔
Magnetic	✘	✔
Resists corrosion	✔	✘
Malleable	✔	✔
Conducts electricity	✔	✔

Iron or aluminium can be used to build cars as they can be pressed into shape; they are both malleable. They are both good electrical conductors.

Aluminium is well-suited to the job because:

- it does not corrode (whereas iron does)
- it is less dense than iron (which means the car will be lighter).

However, iron is well-suited to the job because:

- it is cheaper than aluminium
- it is magnetic (aluminium is not) which means it can be separated for recycling more easily.

However, most cars are made from **steel**. Steel is an alloy of iron and carbon. Steel has different properties from iron which make it more useful. It is harder and does not corrode as fast as iron.

HT Some cars are made from aluminium. Aluminium does not rust or corrode so the car will last for longer. And because aluminium is less dense than steel, the car will:

- be lighter
- get better fuel economy.

Aluminium can be mixed with other metals such as copper and magnesium to create an alloy.

Materials in a Car

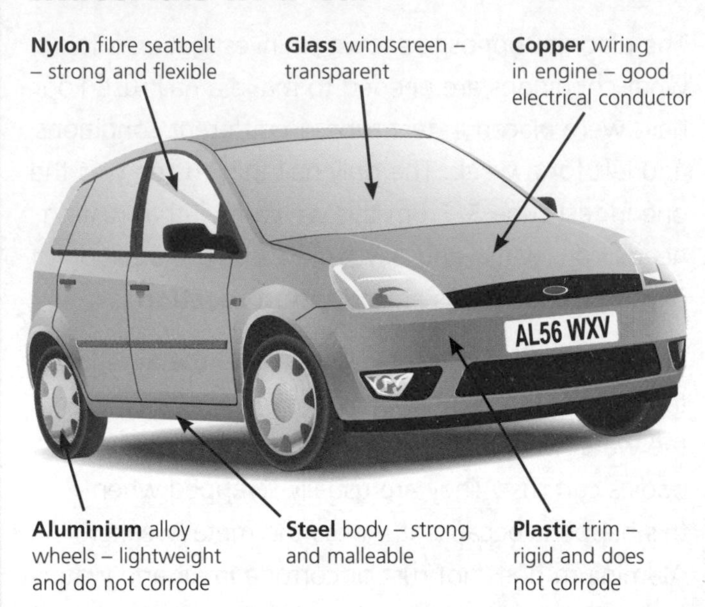

You should be able to suggest what properties are important in a material used in a car, such as the examples in the diagram. If you are given information about the properties of these materials, you should be able to explain why they are used for a particular job in the car.

HT You may be asked make judgements on the suitability of materials used for car manufacture given all the relevant information.

Recycling

Most materials used in a car can be recycled. From 2006, the law requires that 85% of a car must be able to be recycled; this will increase to 95% in 2015.

The problem is separating all the different materials from each other. However, there are many benefits. Recycling materials means:

- fewer disposal problems
- less energy is needed for extracting them from ores
- limited natural resources will last longer.

Recycling the plastics and fibres reduces the amount of crude oil needed to make them, and conserves oil reserves. There are a number of materials in a car, e.g. lead in the car battery, which would cause pollution if put into landfill, so recycling also protects the environment.

Manufacturing Chemicals: Making Ammonia

Ammonia

Ammonia (NH_3) is an alkaline gas. It is made from nitrogen and hydrogen. Getting these gases to combine chemically and stay combined is very difficult. This is because the reaction is reversible: as well as nitrogen and hydrogen combining to form ammonia, the ammonia decomposes in the same conditions to form hydrogen and nitrogen.

Reversible reactions have the symbol $\rightleftharpoons$ in their equation to show that the reaction can take place in either direction.

Ammonia can be used to make nitric acid and fertilisers. Farmers rely on cheap fertilisers made from ammonia to produce enough food for an ever-growing world population.

The Haber Process

Fritz Haber was the first to work out how to make ammonia on a large scale. The raw materials are **nitrogen** (obtained from the air) and **hydrogen** (from natural gas or the cracking of crude oil).

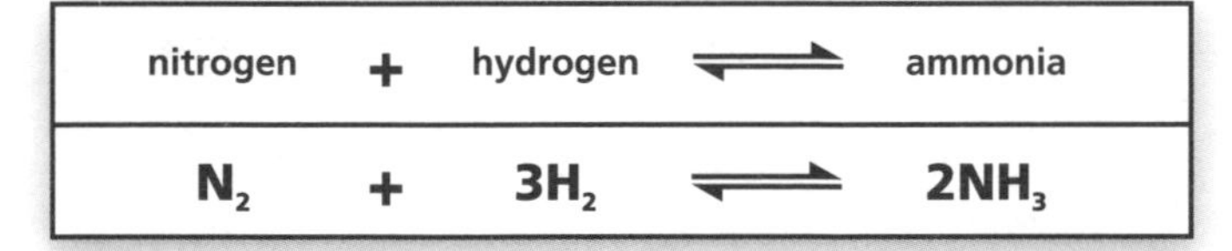

The mixture of 1 part nitrogen and 3 parts hydrogen is compressed to a high pressure and passed into a reactor. The gases are passed over an iron catalyst at 450°C. This is where the reaction takes place (see diagram below).

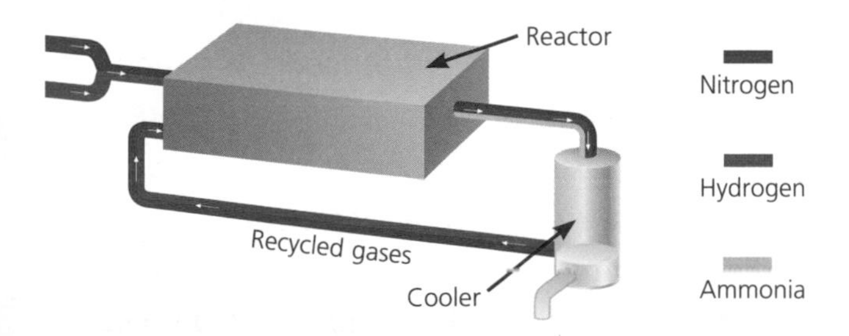

About 28% of the gases are converted into ammonia, which is separated from the unreacted hydrogen and nitrogen by cooling, and collected as a liquid. The unreacted gases are recycled.

You may be asked to interpret graphs and tables containing data about the conditions in the Haber process.

Example

Interpret the graph and table below to explain how temperature and pressure affect the rate of reaction in the Haber process.

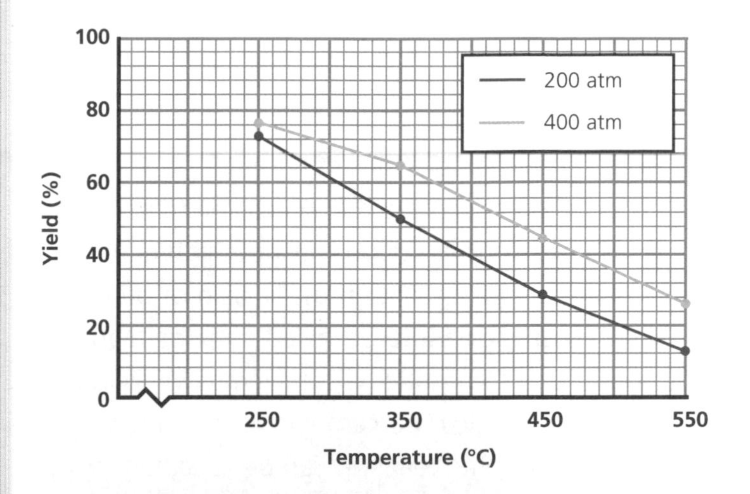

% Yield / Pressure	Temperature			
	250°C	350°C	450°C	550°C
200 atm	73%	50%	28%	13%
400 atm	77%	65%	45%	26%

From the above information you should be able to pick out, for example, that the yield falls when temperature is increased and that the yield increases as pressure increases.

> HT You may also be asked to interpret data on other industrial processes in terms of rate, percentage yield and cost.

Cost

The cost of making a new substance depends on:

- the price of energy (gas and electricity)
- labour costs (wages for employees)
- how quickly the new substance can be made
- the cost of starting materials (reactants)
- the cost of equipment needed (plant and machinery).

Factors Affecting Cost

There are various factors that affect the cost of making a new substance, including:

- the pressure required – the higher the pressure the higher the plant cost
- the temperature required – the higher the temperature the higher the energy cost
- the catalysts required – catalysts reduce costs because they increase the rate of reaction, but they need to be purchased in the first place which increases initial costs
- the number of people required to operate machinery – automation reduces the wage bill
- the amount of unreacted material that can be recycled – recycling reduces costs.

HT

Economic Considerations

Economic considerations determine the conditions used in the manufacture of chemicals:

- The **rate of reaction** must be high enough to produce a sufficient daily yield of product.
- **Percentage yield** achieved must be high enough to produce a sufficient daily yield of product (a low percentage yield is acceptable providing the reaction can be repeated many times with **recycled starting materials**).
- **Optimum conditions** should be used to give the most economical reaction (this could mean producing a slower reaction or a lower percentage yield at a lower cost).

Economics of the Haber Process

$$N_2 + 3H_2 \rightleftharpoons 2NH_3$$

There is great economic importance attached to producing the maximum amount of ammonia in the shortest possible time at a reasonable cost. This demands some compromise.

Effect of Temperature

If a high temperature is used in the Haber process it has two effects:

- It speeds up the rate of the reaction and ammonia is made faster.
- It reduces the percentage yield of ammonia.

The compromise is between making more ammonia slowly at a low temperature and making less ammonia more quickly at a higher temperature.

Effect of Pressure

Using a high pressure in the Haber process gives a higher percentage yield but it is expensive to construct the reaction chamber, and to maintain it to contain such high pressure. The compromise here is between the cost of high pressure and the percentage yield of ammonia.

Effect of a Catalyst

Using a catalyst increases the rate of the reaction, although it does not affect the percentage yield. However, although using a catalyst can reduce costs, purchasing it increases initial costs.

In summary:

- a low temperature increases yield but the reaction is too slow
- a high pressure increases yield but the reaction is too expensive
- a catalyst increases the rate of reaction but does not change the percentage yield.

Therefore a compromise is reached in the Haber process and the following conditions are used:

- An iron catalyst.
- A high pressure of 200 atmospheres.
- A temperature of 450°C which gives a fast reaction with sufficiently high percentage yield.

In your exam, you may be asked to use the above ideas to interpret how rate, cost and yield affect other industrial processes.

Acids and Bases

Acids are substances that contain hydrogen ions (H^+) in solution. pH is a measure of the concentration of H^+ ions in the solution. (Acids have a pH of less than 7.) **Bases** are the oxides and hydroxides of metals. The bases that are soluble are called **alkalis**.

The pH of a solution can be determined by using **universal indicator**. You just need to add a few drops of the solution to the substance and compare the resulting colour to the pH chart as shown below.

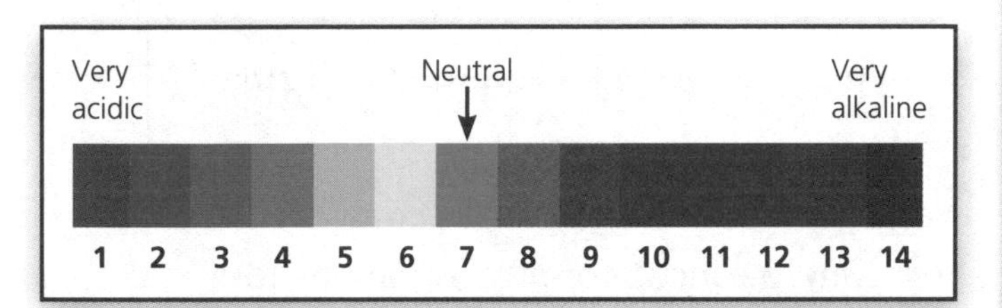

Universal indicator changes colour gradually as the pH changes, whereas some indicators only have one colour change e.g. red-coloured litmus changes suddenly to blue when an alkali is added.

Neutralisation

Acids and bases (alkalis) are chemical opposites. If they are added together in the correct amounts they can cancel each other out. This is called **neutralisation** because the solution that remains has a neutral pH of 7.

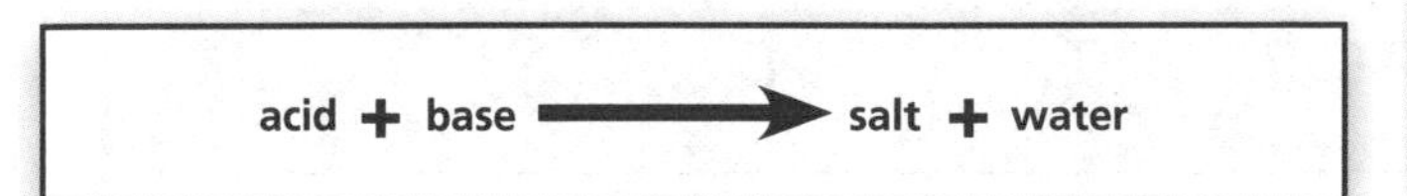

For example, adding hydrochloric acid (HCl) to potassium hydroxide (KOH) is a neutralisation reaction:

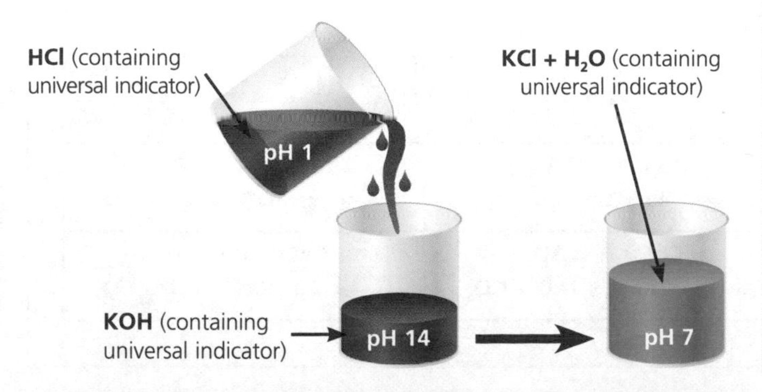

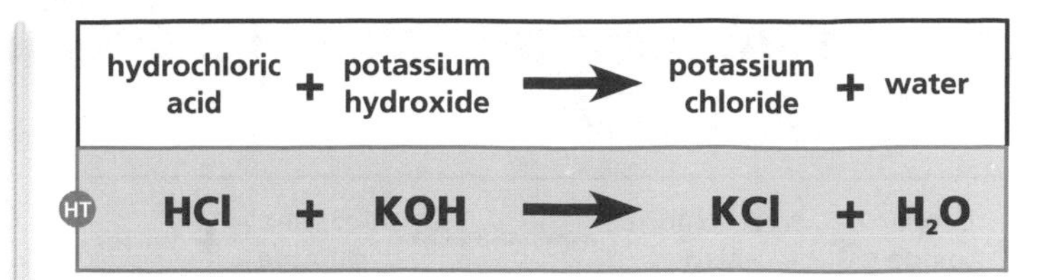

Carbonates **neutralise** acids to produce a salt and water, but they also produce carbon dioxide gas.

Word Equations and Naming Salts

You may be asked to write a word equation for a neutralisation reaction and you may need to work out the name of the **salt** that is produced.

The first name of a salt made by neutralisation comes from the first name of the base or carbonate used, for example:

- **sodium** hydroxide will make a **sodium** salt
- **copper** oxide will make a **copper** salt
- **calcium** carbonate will make a **calcium** salt
- **ammonia** will make an **ammonium** salt.

The second name of the salt comes from the acid used, for example:

- hydro**chlor**ic acid will produce a **chlor**ide salt
- **sulf**uric acid will produce a **sulf**ate salt
- **nitr**ic acid will produce a **nitr**ate salt.
- **phosph**oric acid will produce a **phosph**ate salt.

For example, adding **potassium** hydroxide to **nitr**ic acid to neutralise it will make **potassium nitrate**.

Fit the names of the reactants and products into the general equation. The word equation for this last reaction is:

Word Equations and Naming Salts (cont)

Other examples include:

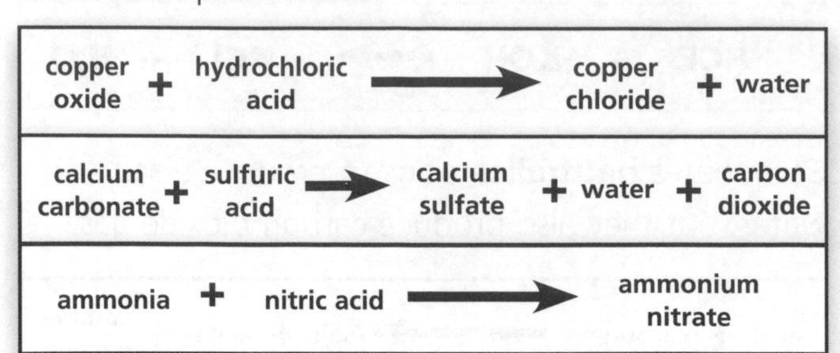

Do not try to use formulae or mix formulae with words if you are asked to write a word equation. Always use + signs and the arrow.

HT Neutralisation Reactions

Neutralisation can be summarised by looking at what happens to the ions in the solutions:

- Alkalis in solution contain **hydroxide ions**, **OH^-**(aq).
- Acids in solution contain **hydrogen ions**, **H^+**(aq).

Neutralisation can therefore be described using the following ionic equation:

$$H^+(aq) + OH^-(aq) \rightleftharpoons H_2O(l)$$

You should be able to construct any of the following balanced formula equations for producing salts:

acid + base → salt + water

	Hydrochloric acid (HCl)	Sulfuric acid (H_2SO_4)	Nitric acid (HNO_3)
Sodium hydroxide (NaOH)	$NaOH + HCl \rightarrow NaCl + H_2O$	$2NaOH + H_2SO_4 \rightarrow Na_2SO_4 + 2H_2O$	$NaOH + HNO_3 \rightarrow NaNO_3 + H_2O$
Potassium hydroxide (KOH)	$KOH + HCl \rightarrow KCl + H_2O$	$2KOH + H_2SO_4 \rightarrow K_2SO_4 + 2H_2O$	$KOH + HNO_3 \rightarrow KNO_3 + H_2O$
Copper(II) oxide (CuO)	$CuO + 2HCl \rightarrow CuCl_2 + H_2O$	$CuO + H_2SO_4 \rightarrow CuSO_4 + H_2O$	$CuO + 2HNO_3 \rightarrow Cu(NO_3)_2 + H_2O$

acid + base → salt

	Hydrochloric acid (HCl)	Sulfuric acid (H_2SO_4)	Nitric acid (HNO_3)
Ammonia (NH_3)	$NH_3 + HCl \rightarrow NH_4Cl$	$2NH_3 + H_2SO_4 \rightarrow (NH_4)_2SO_4$	$NH_3 + HNO_3 \rightarrow NH_4NO_3$

acid + carbonate → salt + water + carbon dioxide

	Hydrochloric acid (HCl)	Sulfuric acid (H_2SO_4)	Nitric acid (HNO_3)
Sodium carbonate (Na_2CO_3)	$Na_2CO_3 + 2HCl \rightarrow 2NaCl + H_2O + CO_2$	$Na_2CO_3 + H_2SO_4 \rightarrow Na_2SO_4 + H_2O + CO_2$	$Na_2CO_3 + 2HNO_3 \rightarrow 2NaNO_3 + H_2O + CO_2$
Calcium carbonate ($CaCO_3$)	$CaCO_3 + 2HCl \rightarrow CaCl_2 + H_2O + CO_2$	$CaCO_3 + H_2SO_4 \rightarrow CaSO_4 + H_2O + CO_2$	$CaCO_3 + 2HNO_3 \rightarrow Ca(NO_3)_2 + H_2O + CO_2$

Fertilisers

Fertilisers are chemicals that farmers use in order to provide their plants with the **essential elements** they need for growth. They increase the crop **yield**. The three main essential elements found in fertilisers are:

- nitrogen, N
- phosphorus, P
- potassium, K.

Fertilisers must be soluble in water so that they can be taken in by the roots of plants.

The following fertilisers can be made by neutralising an acid with an alkali:

- Ammonium nitrate is manufactured by neutralising nitric acid with ammonia.
- Ammonium sulfate is manufactured by neutralising sulfuric acid with ammonia.
- Ammonium phosphate is manufactured by neutralising phosphoric acid with ammonia.
- Potassium nitrate is manufactured by neutralising nitric acid with potassium hydroxide.

Urea made from ammonia can also be used as a fertiliser.

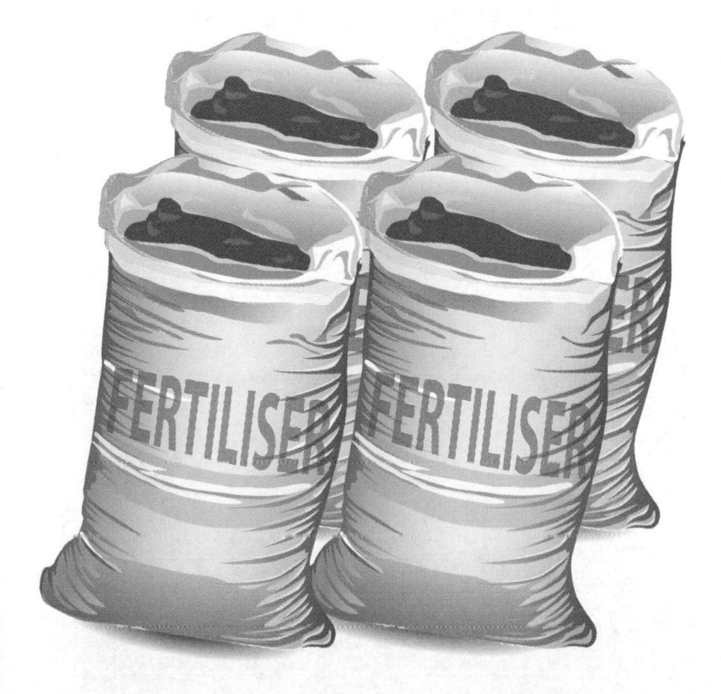

Making a Fertiliser

To make a fertiliser (e.g. potassium nitrate), follow these steps (see diagrams below). Make sure you recognise the apparatus.

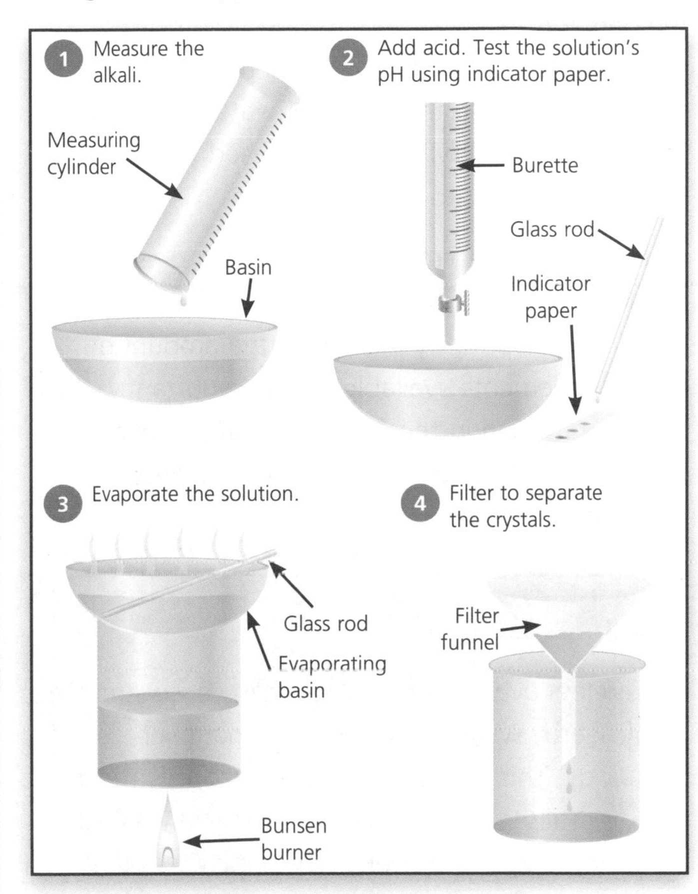

HT

1. Measure out the alkali (e.g. potassium hydroxide) into a basin using a measuring cylinder.
2. Add the acid (e.g. nitric acid) from a burette. Use a glass rod to put a drop of solution onto indicator paper to test the pH. Continue to add the acid a bit at a time until the solution is neutral (pH 7).
3. Evaporate the solution slowly until crystals form on the end of a cold glass rod placed in the solution. Leave to cool and crystallise.
4. Filter to separate the crystals from the solution.
5. Remove the crystals, wash them and leave to dry.

This method is another example of producing a salt (a fertiliser) by neutralisation.

Fertilisers and Crop Yield

Benefits and Problems of Fertilisers

The use of chemical fertilisers helps us grow more food. The world's population is rising, increasing the demand for food. Fertiliser use also causes problems. Chemical fertilisers can pollute water supplies and cause the death of water creatures (**eutrophication**) when they are used too much and are washed into rivers and lakes.

Fertilisers increase crop yield by replacing essential elements in the soil that have been used up by a previous crop, or by increasing the amount of essential elements available. More importantly, they provide nitrogen in the form of **soluble nitrates** which are used by the plant to make protein for growth.

HT Eutrophication

Careless overuse of fertilisers can cause stretches of nearby water to become stagnant very quickly. This process is called **eutrophication**:

1. Fertilisers used by farmers may be washed into lakes and rivers, which increases the level of nitrates and phosphates in the water and increases the growth of simple algae.

2. The algal bloom blocks off sunlight to other plants, causing them to die and rot.

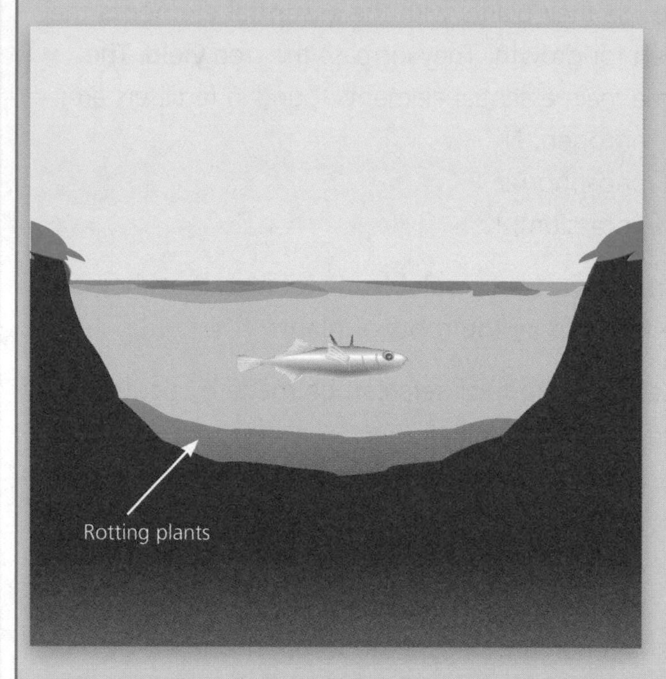

3. This leads to a massive increase in the number of aerobic bacteria (which feed on dead organisms), which quickly use up the oxygen. Eventually, nearly all the oxygen is removed. This means there is not enough left to support the larger organisms, such as fish and other aquatic animals, causing them to suffocate.

Sodium Chloride

Sodium chloride is normal table salt and is used as a flavouring and preservative. It is also a very important raw material for the chemical industry.

In the UK, sodium chloride can be obtained from sea water. It is mined in Cheshire as a solid (rock salt) for gritting roads and also by solution mining for the chemical industry. Salt mining has led to subsidence of the ground in some parts of Cheshire.

The electrolysis of concentrated **sodium chloride solution** (also known as salt solution or brine) forms hydrogen at the **cathode** (negative electrode) and chlorine at the **anode** (positive electrode). Sodium hydroxide is also formed in the solution.

The electrodes must be made from **inert** materials as the products are very reactive.

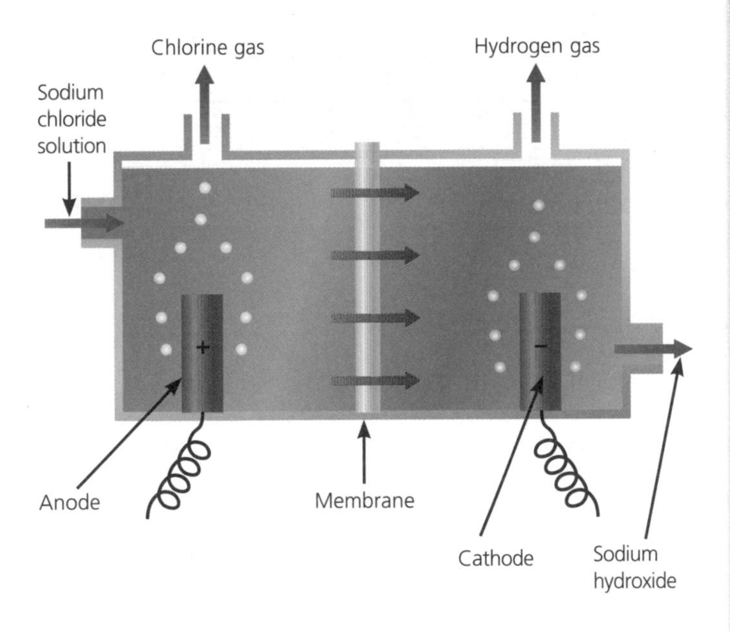

There are many uses of the products obtained from the electrolysis of sodium chloride:

- Hydrogen is used to make margarine.
- Sodium hydroxide is used to make soap.
- Chlorine is used to sterilise water, make solvents and make plastics, for example, PVC.
- Chlorine and sodium hydroxide are reacted together to make household bleach.

The test for chlorine is that it bleaches moist litmus paper.

Test for Chlorine

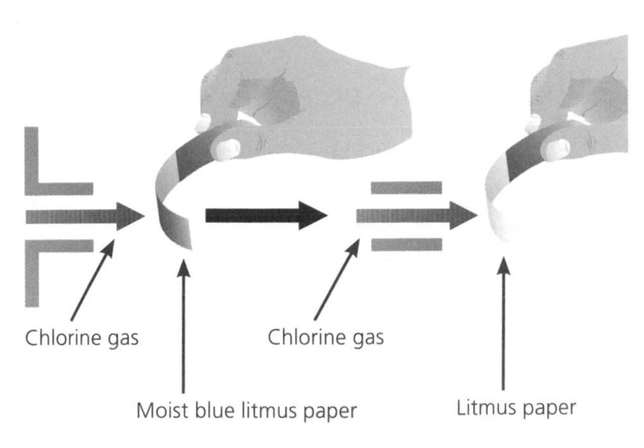

HT Electrolysis of Sodium Chloride Solution

Sodium chloride solution contains the ions Na^+, H^+, Cl^- and OH^-. The H^+ and Cl^- ions are discharged at the electrodes.

Half-equations can be written for the reactions that take place at the electrodes in the electrolysis of concentrated sodium chloride solution.

- Hydrogen is made at the cathode:

$$2H^+ + 2e^- \longrightarrow H_2$$

Electrons are gained – this is reduction.

- Chlorine is made at the anode:

$$2Cl^- - 2e \longrightarrow Cl_2$$

Electrons are lost – this is oxidation.

Sodium ions and hydroxide ions are not discharged and remain to make sodium hydroxide solution.

The electrolysis of sodium chloride is a very important part of the chemical industry; many other parts of industry depend on the products manufactured by this process.

Exam Practice Questions

1 **a)** Explain why some people live near volcanoes. **[1]**

b) What is the lithosphere? **[1]**

c) Explain why the plate tectonic theory is now accepted by most scientists. **[2]**

2 **a)** Quarries take up land and destroy the landscape. Outline what other problems there are with this method of extracting rock from the ground. **[3]**

b) Limestone thermally decomposes into calcium oxide (CaO) and carbon dioxide (CO_2). Write a balanced symbol equation for the thermal decomposition of limestone, $CaCO_3$. **[1]**

3 **a)** Steel is an alloy composed mainly of iron. Explain why steel is more useful than iron. **[2]**

b) Name the metal that is mixed with lead to make solder. **[1]**

c) Give two properties that make solder useful for welding metal gas pipes together. **[2]**

4 Look at the data table. It shows how temperature and pressure affect the yield (%) of ammonia in the Haber process.

	Temperature (°C)			
Pressure (atmospheres)	**250**	**350**	**450**	**550**
50	60%	30%	11%	4%
100	67%	34%	16%	7%
200	73%	50%	29%	14%

a) What temperature gives the lowest yield? **[1]**

b) What is the yield at 100 atmospheres pressure and a temperature of 350°C? **[1]**

c) What happens to the yield as the temperature is decreased? **[1]**

5 Explain why ammonium phosphate, $(NH_4)_3PO_4$, can be used as a fertiliser. **[3]**

6 **a)** Name the three important products made when sodium chloride solution undergoes electrolysis. **[3]**

b) Which two materials are reacted together to make household bleach? **[2]**

c) Complete and balance the ionic equation for the reaction at the anode during the electrolysis of sodium chloride solution. **[2]**

$Cl^- - \_\_\_\_ \longrightarrow Cl_2$

HT 7 Describe what happens when a continental plate and an oceanic plate collide. **[3]**

8 Explain why granite, limestone and marble have different levels of hardness. **[3]**

9 Describe the conditions used in the Haber process and explain why these conditions are selected. **[6]**

The quality of your written communication will be assessed in your answer to this question.

10 Describe the problems caused by the use of too much chemical fertiliser near rivers and lakes. **[3]**

11 Nitric acid (HNO_3) reacts with sodium carbonate (Na_2CO_3) to make sodium nitrate ($NaNO_3$), water (H_2O) and carbon dioxide (CO_2). Write a balanced symbol equation for this reaction. **[2]**

C3: Chemical Economics

This module looks at:
- How rates of reaction can be investigated.
- How temperature and concentration affect the rate of reaction.
- How surface area and catalysts affect the rate of reaction.
- Calculating relative formula mass, and calculating masses in reactions.
- Percentage yield and atom economy calculations.
- Exothermic and endothermic reactions; measuring reaction energy changes.
- Batch and continuous processes, manufacturing pharmaceuticals.
- Properties and uses of diamond, graphite, buckminster fullerene and nanotubes.

Rates of Reaction

Chemical reactions occur at different rates: some are very slow, such as rusting, and some, for example burning and explosions, are very fast.

We can measure the time a reaction takes. Here are some times taken for a 2cm strip of magnesium to completely dissolve in different concentrations of an acid:

Concentration	Reaction Time (s)
0.01	205
0.02	114
0.03	72
0.04	33
0.05	17

You may be asked to plot results such as these on a graph where the axes have been prepared for you. You might be asked to interpret this information, for example, pick out the fastest reaction from this table or predict the time it would take if the acid concentration was 0.035 (i.e. about half way between 33 and 72, so 53 seconds).

The reaction stops because one of the **reactants** is all used up. In this case the magnesium is the **limiting reactant**, the one that is used up; the acid is in excess as some is left over at the end of the reaction. The amount of **product** is increased if more reactant is used; it is directly proportional to the amount of reactant, i.e. if you double the amount of reactant then the amount of product also doubles.

HT The amount of the limiting reactant you start with dictates the amount of product you will make. When there are more reactant particles, there is more reaction and more product particles are produced.

Gas Reactions

When a gas is made you can follow the progress of the reaction by recording the mass lost as the gas escapes, or by collecting the gas and measuring the volume:

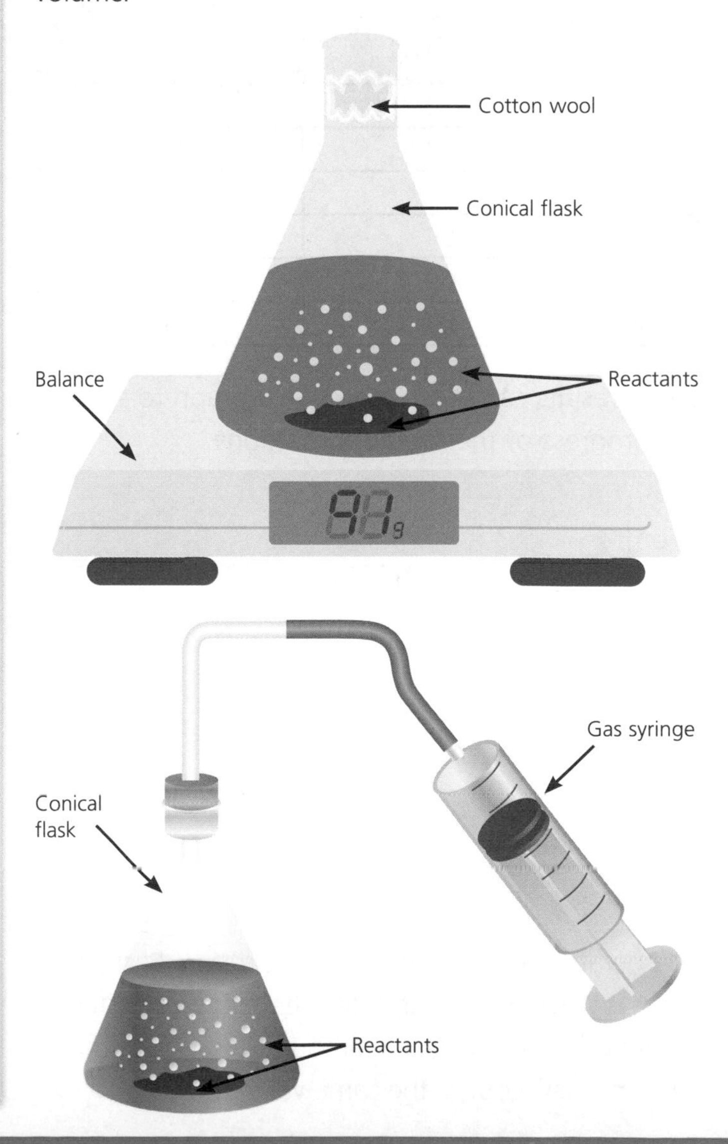

Gas Reactions (cont)

Measuring the volume of gas made in a particular time would give you the rate of the reaction, or the fastest reaction would fill the syringe in the shortest time.

HT The rate of a reaction can be calculated to give the average volume of gas made in a particular time. The unit would be cm^3/s or cm^3/min. Alternatively, if mass is used the units would be g/s or g/min.

Analysing the Rate of Reaction

Two reactions were carried out at different temperatures and the following results were obtained:

Time (s)	Volume of gas made at temperature A (cm^3)	Volume of gas made at temperature B (cm^3)
0	0	0
30	22	18
60	40	35
90	54	49
150	62	57
180	62	62

These results can be sketched on a graph to show the progress of the chemical reactions.

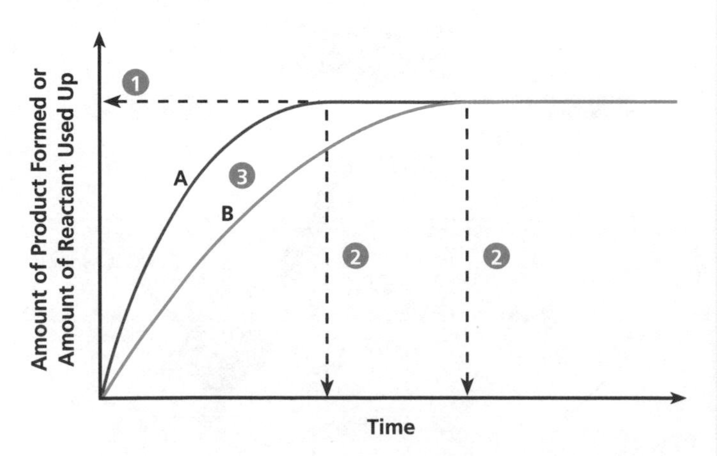

Temperature A is hotter than temperature B: the line for A is steeper showing that the reaction is faster. They both make the same amount of product; the two lines level out at the same value.

The graphs for concentration and pressure would be very similar with line A representing the higher concentration or the higher pressure.

You may be asked to plot results such as these on a graph where the axes have been prepared for you.

For example:

1. **Show how much product was made** by drawing a horizontal line from the highest point on the graph across to the y-axis.
2. **Show how long it takes to make the products.** The flat line on the graph indicates that the reaction is finished and that the products have been made. By drawing a vertical line down to the x-axis (time) from the flat line we can see how long this took.
3. **See which reaction is quicker** by comparing the steepness of the lines (the steeper the line, the quicker the reaction).

HT The graph below shows the progress of a chemical reaction. The points used to plot the graph are shown as crosses.

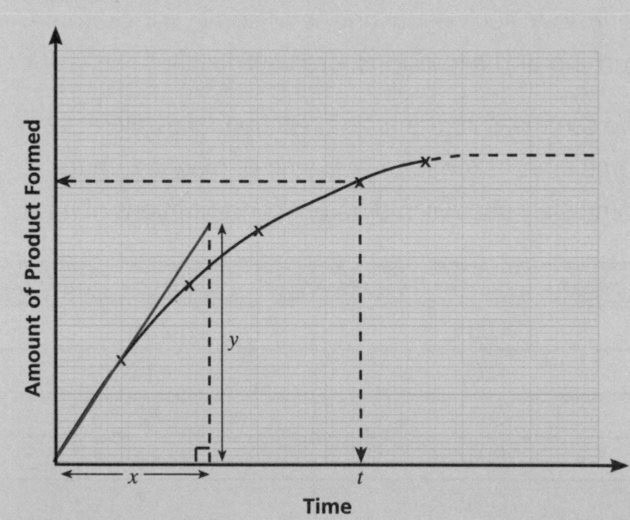

From the graph you can:

- calculate the initial rate of reaction by drawing a straight line following the start of the curve and working out $\frac{y}{x}$
- extend the curve by estimating the most likely path it will take next
- work out the amount of product formed by a time (t) for which we did not have a reading.

Collisions and Rate of Reaction

The particles in a chemical reaction must collide together for a reaction to take place. The more collisions there are, the faster the reaction. The idea of collisions is used to explain how different factors affect the rate of a reaction.

Temperature of the Reactants

Lower temperature – lower rate	Higher temperature – higher rate
In a cold reaction mixture, the particles move quite slowly. The particles will collide with each other less often and with less energy, so fewer collisions will be successful. 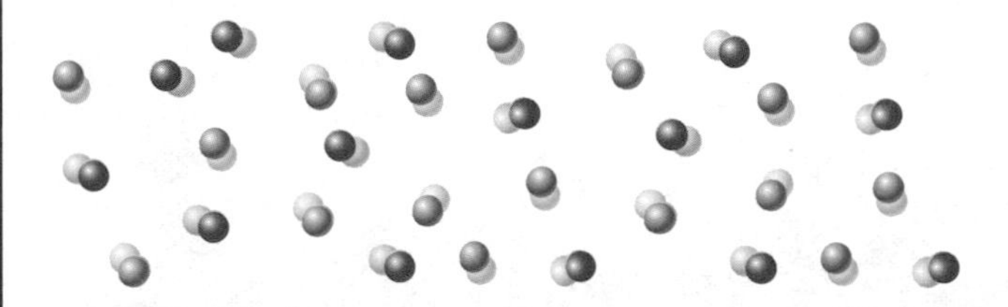	If the temperature of the reaction mixture is increased, the particles will move faster. They will collide with each other more often and with greater energy, so many more collisions will be successful. 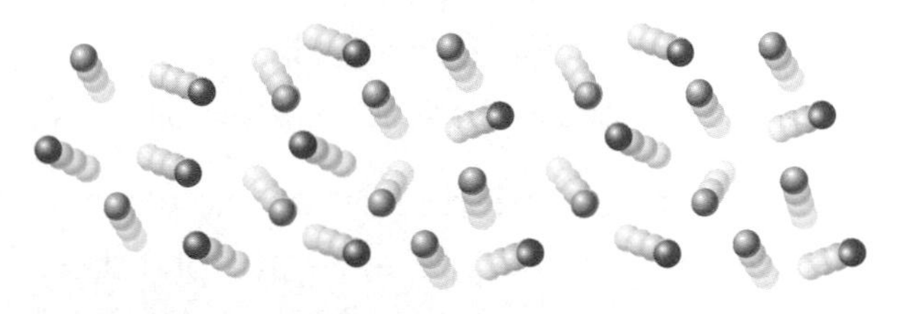

Concentration of the Reactants

Lower concentration – lower rate	Higher concentration – higher rate
In a reaction where one or both reactants are in low **concentrations**, the particles are spread out. The particles will collide with each other less often, resulting in fewer successful collisions. 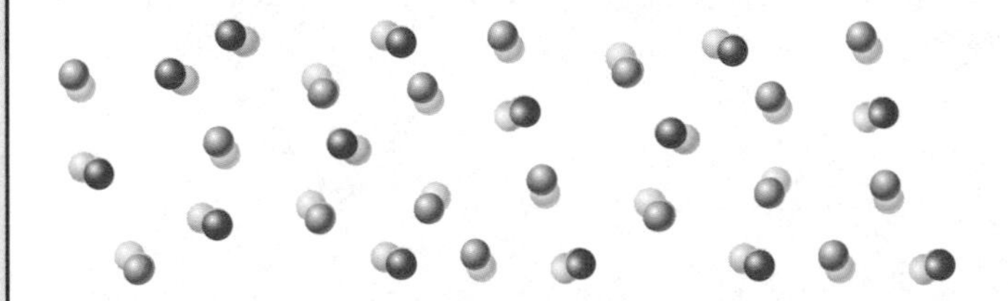	Where there are high concentrations of one or both reactants, the particles are crowded close together. The particles will collide with each other more often, resulting in many more successful collisions. 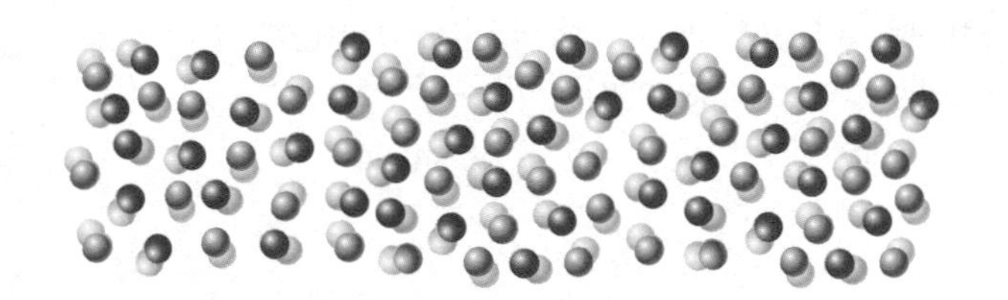

Pressure of a Gas

Lower pressure – lower rate	Higher pressure – higher rate
When a gas is under a low pressure, the particles are spread out. The particles will collide with each other less often resulting in fewer successful collisions. (This is like low concentration of liquid reactants.) 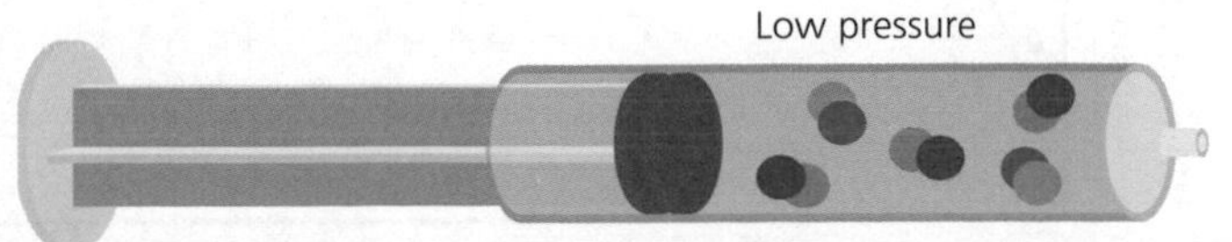	When the pressure is high, the particles are crowded more closely together. The particles collide more often, resulting in many more successful collisions. (This is like high concentration of liquid reactants.) 

C3 Rate of Reaction (2)

HT Collision Theory

Increasing **temperature** causes an increase in the kinetic energy of the particles, i.e. they move a lot faster. The faster the particles move, the greater the chance of them colliding, so the number of collisions per second increases. The more collisions there are between particles, the faster the reaction.

When the particles collide at an increased temperature they have more energy. When a collision has more energy, the chance of it causing a successful collision is increased (energetic collisions = more successful collisions).

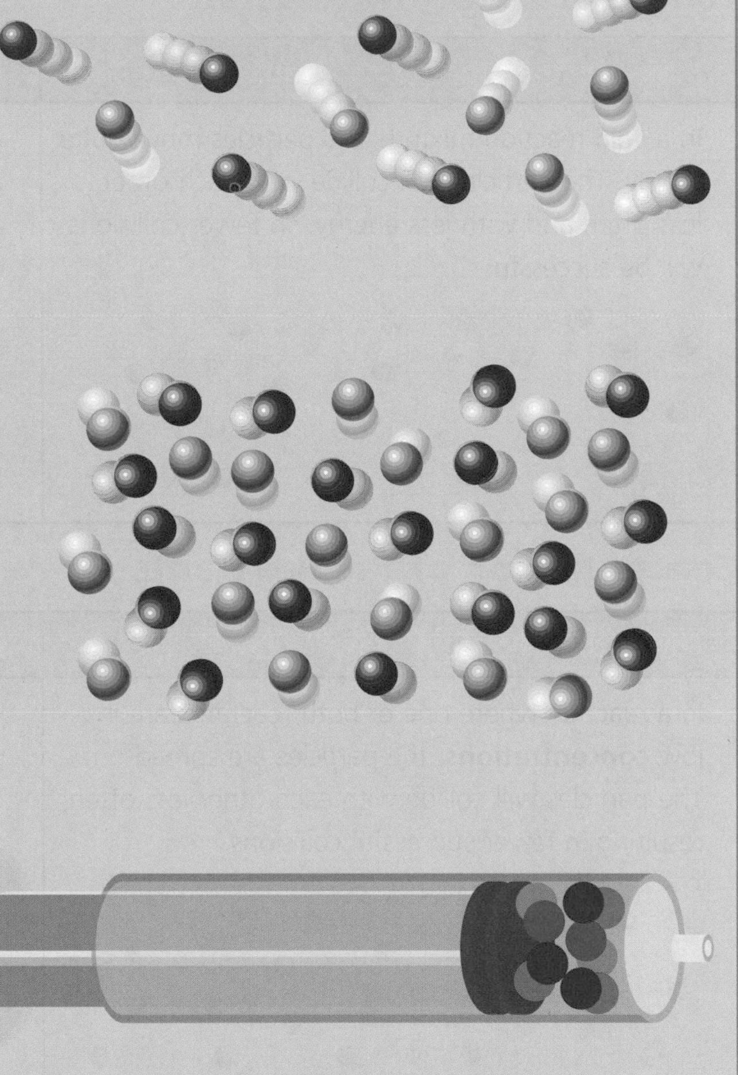

Increasing **concentration** increases the number of particles in the same space, i.e. the particles are much more crowded together.

Increasing **pressure** in a gas reaction is very much like increasing the concentration; the particles are more crowded together and this increases the frequency of collisions between the particles.

The more crowded the particles are, the greater the chance of them colliding together, which increases the number of collisions per second. (*More frequent* collisions not just *more* collisions.)

You may be asked to interpret information about the effect of changing temperature, concentration or pressure on a reaction. This could be in tables, graphs or a description. Use the guidelines on page 40 to help.

You may also be asked to sketch a graph showing the effect of changing these variables. This is a sketch graph showing the effect of temperature – temperature A is higher than temperature B:

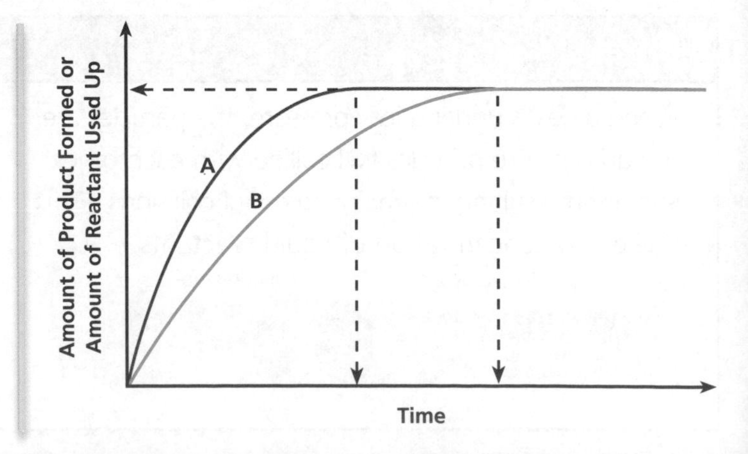

Surface Area of Solid Reactants

Powdered solids react faster than lumps of the same reactant. A powdered reagent has a much larger surface area compared to lumps of the same material. There are more particles available on the surface for the other reactants to collide with. The greater the number of particles exposed, the greater the chance of them colliding together, which increases the reaction. (More collisions = faster reaction.)

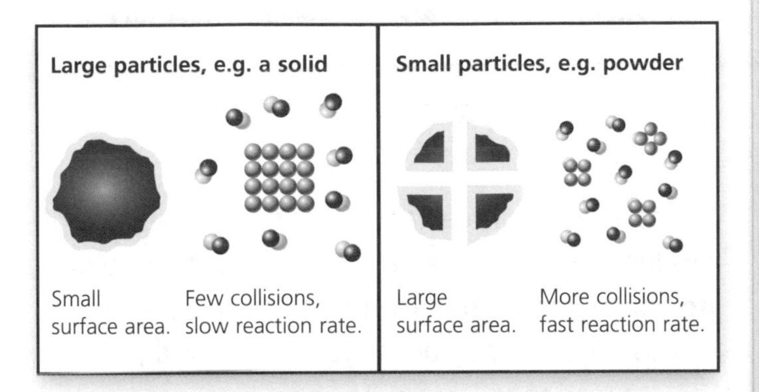

Gas and fine dust have the largest surface area of all and sometimes they react so fast it leads to an explosion, e.g. burning hydrogen or a custard powder explosion. An **explosion** is a very fast reaction where huge volumes of gas are made. Other materials that can explode are dynamite and TNT.

Factories that handle powders such as flour, custard powder or sulfur have to be very careful because the dust of these materials can mix with air and could cause an explosion if there is a spark. The factories have to prevent dust being produced and take precautions to ensure no spark is made that would ignite a dust–air mixture.

Using a Catalyst

A **catalyst** is a substance that increases the rate of a reaction and is unchanged at the end of the reaction. Catalysts are very useful materials, as only a small amount of catalyst is needed to speed up the reaction of large amounts of reactant.

HT Catalysts (like reactants) are most effective when they have a large surface area. The greater the number of particles exposed, the greater the chance of them colliding together, which means the number of collisions per second increases. (*More frequent* collisions, not just *more* collisions.)

You may be asked to sketch a graph showing the effect of using a catalyst in an experiment, such as this:

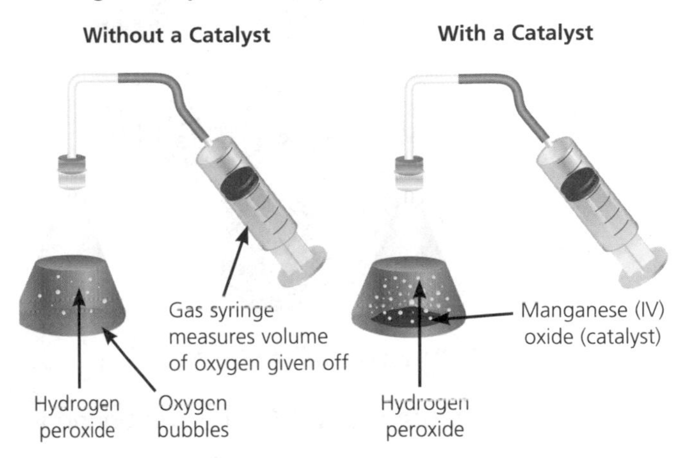

The sketch graph would look like this:

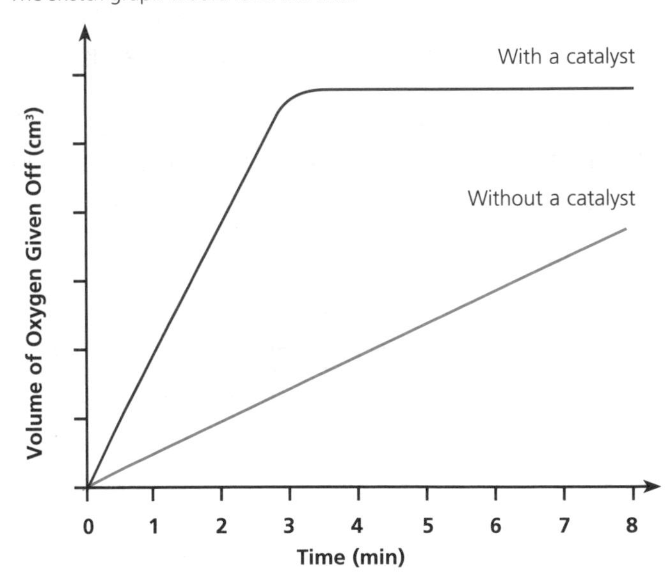

The sketch graph for changing the surface area would look very similar with the steeper line representing the larger surface area. You may be asked to interpret information about the effect of changing the surface area or using a catalyst on a reaction. This data could be in tables, graphs or a description.

Relative Atomic Mass, A_r

Atoms are too small for their actual atomic mass to be of much use to us. A more useful measure is **relative atomic mass, A_r**. Use the Periodic Table to look up the relative atomic mass.

Each **element** in the Periodic Table has two numbers. The larger of the two numbers (top left below) is the A_r of the element.

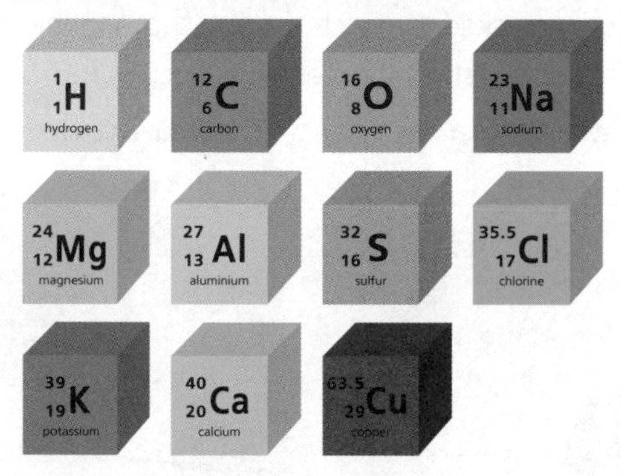

Relative Formula Mass, M_r

The **relative formula mass, M_r**, of a compound is simply the A_rs of all its elements added together. To calculate the M_r, the formula of the compound and the A_r of all the atoms involved are needed.

There are a number of ways to set out an M_r calculation. Whichever method you choose, you should always show your working. The simplest method is shown below:

Example 1

Calculate the relative formula mass of H_2SO_4.

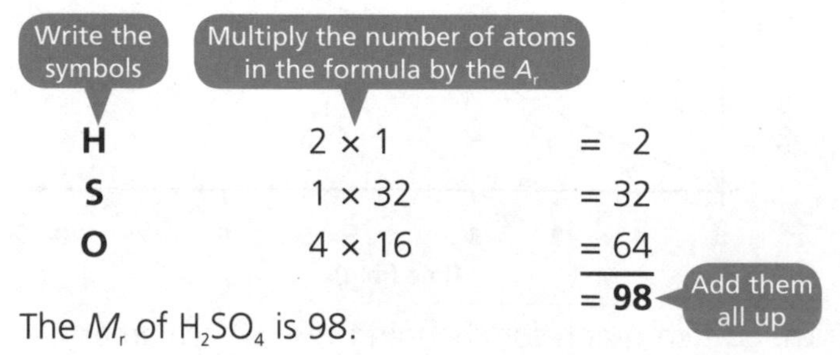

The M_r of H_2SO_4 is 98.

Example 2

Calculate the relative formula mass of $Ca(OH)_2$.

Ca	1 × 40	= 40
O	2 × 16	= 32
H	2 × 1	= 2
		= 74

Conservation of Mass

The total mass of the starting materials (reactants) always equals the total mass of the substances produced (products). This is called the **principle of conservation of mass**. We can show this with a reaction such as when nitric acid (M_r = 63) reacts with ammonia (M_r = 17) to make ammonium nitrate (M_r = 80):

$$HNO_3 + NH_3 \longrightarrow NH_4NO_3$$

Total mass of reactants = 63 + 17 = 80
Total mass of product = 80

The total mass of reactants equals the total mass of products because when a chemical reaction occurs **no atoms are gained or lost**. You end up with exactly the same number as you started with; they are just rearranged into different substances.

Atoms are not created or destroyed during a chemical reaction. You can show that mass is conserved in any reaction, for example, burning propane:

$$C_3H_8 + 5O_2 \longrightarrow 3CO_2 + 4H_2O$$

Calculate the M_r for each reactant and product:

C_3H_8 (3 × 12) + (8 × 1) = 44
O_2 (2 × 16) = 32

CO_2 12 + (2 × 16) = 44
H_2O (2 × 1) + 16 = 18

Multiply by the number of molecules of each and then add the masses together:

Reactants: 44 + (5 × 32) = 204

Products: (3 × 44) + (4 × 18) = 204

Mass of reactants = Mass of products

The idea that the total mass of reactants is the same as the total mass of products in a reaction can be used to calculate how much product is made or how much of a reactant is needed.

Example 1
When 20g of calcium carbonate is heated until it decomposes, 11.2g of calcium oxide is made. We can now work out the mass of carbon dioxide made:

20g of reactant must make 20g of products
20g = 11.2g + mass of carbon dioxide

So the mass of carbon dioxide made
= 20g – 11.2g = **8.8g**

Example 2
What mass of oxygen is needed to react with 4.8g of magnesium to make 8g of magnesium oxide?

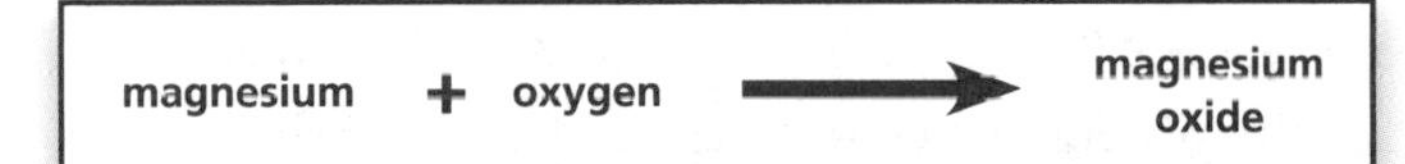

8g of reactants must make 8g of products
4.8g + mass of oxygen = 8g

So the mass of oxygen = 8g – 4.8g = **3.2g**

If you use more reactants then you will get more products. If you use twice as much reactant you get twice as much product. The mass of product obtained is directly proportional to the amount of reactant you start with. This idea can be used to calculate the mass of a reactant or product if you know the reacting ratios.

Example 3
From the previous example we know that 4.8g of magnesium makes 8g of magnesium oxide. We can double it:

2 × 4.8g = 9.6g of magnesium will make
2 × 8g = 16g of magnesium oxide

We can multiply by 10:
10 × 4.8g = 48g of magnesium will make
10 × 8g = 80g of magnesium oxide

Example 4
What mass of magnesium is needed to make 2g of magnesium oxide?

2g of magnesium is $\frac{2g}{8g} = \frac{1}{4}$ of the ratio.
$\frac{1}{4}$ of 4.8g = 1.2g, i.e. 1.2g of magnesium makes 2g of magnesium oxide.

More Calculations

We sometimes need to be able to work out how much of a substance is used up or produced in a chemical reaction. To do this we need to know:

- the relative formula mass, M_r, of the reactants and products (or the relative atomic mass, A_r, of all the elements)
- the balanced symbol equation for the reaction.

By substituting the M_rs into the balanced equation, we can work out the ratio of mass of reactant to mass of product and apply this to the question.

Example 1
When calcium carbonate ($CaCO_3$) is heated, it produces calcium oxide (CaO) and carbon dioxide (CO_2). How much calcium oxide can be produced from 50kg of calcium carbonate?

(Relative atomic masses: Ca = 40, C = 12, O = 16.)

Write down the equation.

$$CaCO_3(s) \xrightarrow{heat} CaO(s) + CO_2(g)$$

Work out the M_r of each substance.

$$40 + 12 + (3 \times 16) \rightarrow (40 + 16) + [12 + (2 \times 16)]$$

Check that the total mass of reactants equals the total mass of the products. If they are not the same, check your work.

$$100 \rightarrow 56 + 44$$ ✔

Since the question only mentions calcium oxide and calcium carbonate, you can now ignore the carbon dioxide. This gives the ratio of mass of reactants to mass of products.

100 : 56

Apply this ratio to the question:

If 100kg of $CaCO_3$ produces 56kg of CaO,
then 1kg of $CaCO_3$ produces $\frac{56}{100}$ kg of CaO
and 50kg of $CaCO_3$ produces $\frac{56}{100} \times 50 =$ **28kg** of CaO.

Percentage Yield and Atom Economy

Percentage Yield

Whenever a reaction takes place, the starting materials (i.e. the reactants) produce new substances (i.e. the products). The greater the amount of reactants used, the greater the amount of products formed.

Percentage yield is a way of comparing the actual amount of product made (the actual yield) to the amount of product theoretically expected, which is the predicted yield. It is calculated using the following formula:

$$\text{Percentage yield} = \frac{\text{Actual yield}}{\text{Predicted yield}} \times 100$$

- A 100% yield means that no product has been lost, i.e. the actual yield is the same as the predicted yield.
- A 0% yield means that no product has been made, i.e. the actual yield is zero.

Example 1

A reaction was carried out to produce the **salt** magnesium sulfate. The correct amounts of reactants were added and allowed to react. The resulting solution was evaporated and the salt crystals were obtained by filtration. The predicted yield of magnesium sulfate was 7g but the actual yield was 4.9g.

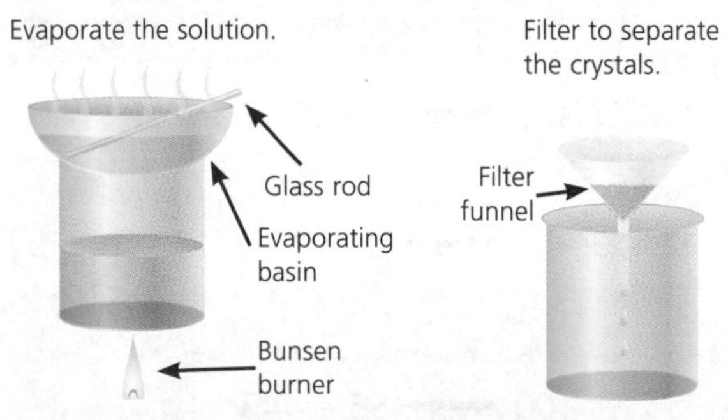

Calculate the percentage yield of magnesium sulfate.

$$\text{Percentage yield} = \frac{\text{Actual yield}}{\text{Predicted yield}} \times 100$$

$$= \frac{4.9\text{g}}{7\text{g}} \times 100 = \mathbf{70\%}$$

The percentage yield of magnesium sulfate was 70% which means that some magnesium sulfate was lost during the process. It could have been lost during evaporation, filtration, the transfer of liquids and / or heating.

Atom Economy

Calculating the **atom economy** is an alternative method of deciding how effective a reaction is. The atom economy is a way of measuring the amount of atoms that are wasted when a chemical is manufactured. It is calculated using the following formula:

$$\text{Atom economy} = \frac{M_r \text{ of the desired products}}{\text{Total } M_r \text{ of all products}} \times 100$$

A 100% atom economy means that all the reactant atoms have been converted into desired product.

The higher the atom economy, the 'greener' the process (less waste).

Example 1

Limestone ($CaCO_3$, $M_r = 100$) is heated to make the useful product calcium oxide (CaO, $M_r = 56$). The other product is carbon dioxide (CO_2, $M_r = 44$). What is the atom economy for this important industrial process?

$$\text{Atom economy} = \frac{M_r \text{ of the desired products}}{\text{Total } M_r \text{ of all products}} \times 100$$

$$= \frac{56}{56 + 44} \times 100 = \mathbf{56\%}$$

You may be asked to make judgements about reactions such as this; with a 56% atom economy this is quite a wasteful process and not very green. A typical yield for the process is 95% and you would then decide that the actual reaction is quite effective.

HT An industrial process needs as high a percentage yield as possible in order to:

- reduce costs
- minimise wasting reactants.

The atom economy also needs to be as high as possible in order to:

- reduce unwanted products
- make the process more sustainable.

Exothermic and Endothermic Reactions

Many reactions are accompanied by a **temperature rise**. These are known as **exothermic** reactions because heat energy is **given out** to the surroundings.

Some reactions are accompanied by a **fall in temperature**. These reactions are known as **endothermic** reactions because heat energy is **taken in** from the surroundings (the reaction absorbs energy).

Comparing Fuels

The equipment shown in the diagram below can be used to compare the amounts of heat energy released by the combustion of different fuels.

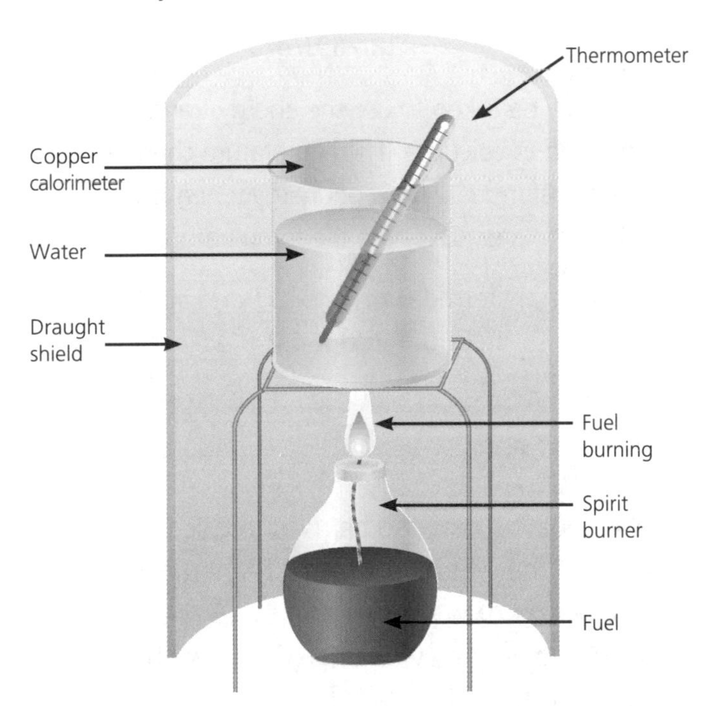

The greater the rise in the temperature of the water, the greater the amount of energy, in joules (J) or kilojoules (kJ), released from the fuel being used.

To make meaningful comparisons we would need to carry out a **fair test** each time. We would need to:

- use the same mass (volume) of water
- use the same calorimeter
- have the burner and calorimeter the same distance apart
- burn the same mass of fuel.

The formula used to work out the change in temperature (°C) is:

Temperature change (°C) = **Final temperature of water** (°C) − **Start temperature of water** (°C)

If you burn the same mass of each fuel, the fuel that produces the largest temperature rise releases the most energy.

1g of three different fuels were burned in the above apparatus. Here are the results:

Fuel	Start Temp. °C	Final Temp. °C	Temperature change °C
Propanol	19	32	13
Ethanol	19	30	11
Pentane	18	34	

You may be asked to calculate the missing temperature change (i.e. 34°C – 18°C = 16°C) and which fuel gave out the most energy (i.e. pentane because it had the largest temperature change).

Breaking and Making Bonds

In a chemical reaction:

- breaking bonds is an **endothermic** process
- making bonds is an **exothermic** process.

HT Chemical reactions that require more energy to break bonds than is released when new bonds are made are **endothermic** reactions.

Chemical reactions that release more energy when bonds are made than is required to break bonds are **exothermic** reactions.

For example, bottled gas (propane) burns in air to release lots of energy. The displayed formula equation for the burning of propane is shown on the next page.

Breaking and Making Bonds (cont)

All the bonds in propane and oxygen have to be broken. Energy needs to be taken in to break the bonds.

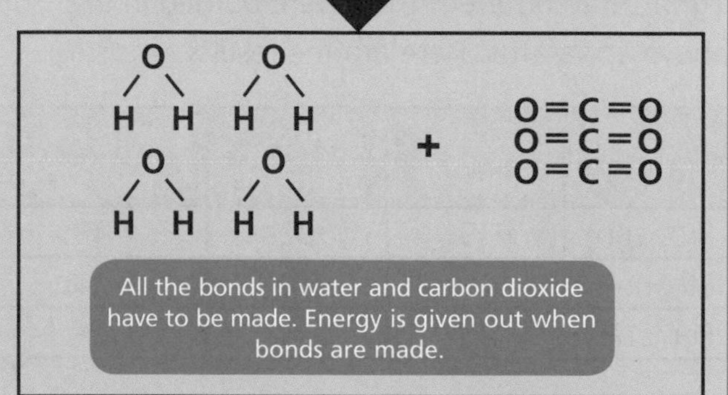

When propane burns, more energy is given out when the new bonds are made than is taken in to break the old bonds at the start. The **overall energy change is exothermic**.

Calculating Energy Changes

In order to compare fuels, we need to work out the amount of energy transferred by the fuel to the water, and the amount of energy transferred per gram of fuel burned. The results from an experiment with hexane are as follows:

	Start	End
Mass of burner and hexane	187.60g	187.34g
Temperature of water	22°C	34°C

Mass of hexane burned (187.60 – 187.34) = **0.26g**

Rise in temperature of water (34°C – 22°C) = **12°C**

Mass of water in calorimeter = **200g**

The amount of energy transferred to the water can be calculated using the following formula:

Energy supplied to warm the water = Mass of water × Specific heat capacity × Temperature change

Energy transferred = $m \times c \times \Delta T$

N.B. You do not need to remember this formula.

Energy supplied = 200g × 4.2J/g/°C × 12°C
= **10 080 joules**

N.B. Specific heat capacity is a constant that is specific to a particular liquid. For water it has a value of 4.2J/g/°C.

The energy transferred per gram of fuel burned can be calculated using the following formula:

$$\textbf{Energy per gram} = \frac{\textbf{Energy supplied}}{\textbf{Mass of fuel burned}}$$

N.B. You need to remember this formula.

$$\text{Energy per gram} = \frac{10\,080}{0.26\text{g}}$$
$$= \mathbf{38\,769J/g}$$

You may be asked to use the energy transferred equation to calculate the temperature change. A fuel transferred 12 600J of energy when heating 200g of water. Calculate the temperature change.

Energy transferred = $m \times c \times \Delta T$ and so
$12\,600 = 200 \times 4.2 \times \Delta T$

Rearrange the equation.

$$\Delta T = \frac{12\,600}{200 \times 4.2} = 15°C$$

The equation can also be used to calculate the mass of the water heated up. The temperature of a beaker of water went up by 18°C when 11 340J of energy was used. Calculate the mass of the water.

Energy transferred = $m \times c \times \Delta T$ and so
$11\,340 = m \times 4.2 \times 18$

Rearrange the equation.

$$m = \frac{11\,340}{4.2 \times 18} = 150\text{g}$$

Batch and Continuous

In a **batch process** the reactants are put into a reactor, the reaction happens and then the product is removed. Medicines and pharmaceutical drugs are often made in batches. Batch processes make a product on demand, make a product on a small scale, can be used to make a variety of products and are more labour intensive because the reactor needs to be filled, emptied and cleaned.

In a **continuous process**, e.g. making ammonia, reactants are continually being fed into a large reactor and the product is continually being produced at the same time (like a conveyer belt). Continuous processes operate all the time and run automatically, make a product on a large scale and are dedicated to just one product.

Making Medicines

The materials used to make a medicine can be **manufactured** (synthetic) or they can be extracted from **natural sources** such as plants. The steps needed to extract a small amount of material from a plant source are as follows:

1. **Crushing** – the plant material is crushed using a mortar and pestle.
2. **Dissolving** – a suitable solvent is added to dissolve the material. This could be **boiled** to improve the extraction.
3. **Chromatography** – a concentrated solution of the material is spotted onto chromatography paper and allowed to separate.

Developing Medicines

It takes a long time – over 10 years – from discovering a material that will act as a medicine, to being able to use it on patients. **Research** needs to be carried out into new pharmaceutical materials. This cannot be automated (carried out by machines) as decisions need to be made, so highly qualified scientists are needed. This means that labour costs are high. Further research is then carried out to **develop** the drug to increase its effectiveness before it is **tested** to ensure it works properly, is safe to use and has no serious side-effects. The medicine must then be approved for use and must satisfy all the **legal requirements** before it can be sold.

Medicines are expensive because the materials could be rare or may require complex methods to extract the **raw materials** (starting materials) from plants. Medicines are made in small quantities and it is not possible to totally automate the manufacturing process. Medicines must be as pure as possible to avoid side effects caused by impurities. Making medicines is labour intensive and, therefore, staff costs are high. The marketing of a new medicine is also very expensive.

HT The research into, and development of, a new pharmaceutical material takes a few years. Hundreds of similar molecules have to be made and tested to find the one that works best. Sometimes the best molecules have too many or serious side effects, so more development is needed. Once a material is developed as a medicine it has to undergo lots of testing to ensure it is better than other available medicines and is safe to use. There are many regulations on how it can be tested, such as when, how and if it can be tested on animals. Ultimately it has to be tested on human volunteers. Some countries have very strict legal rules that a new medicine must satisfy before it can be put on the market.

The table is a summary of how medicines can be tested to make sure they are pure:

Test	Result if Pure	Result if Impure
Melting point	Single, sharp temperature.	Melts over range of temperatures.
Boiling point	Single, sharp temperature.	Boils over range of temperatures.
Chromatography	Single dot on paper.	Two or more dots on paper.

You may be given some information about a medicine and be asked to decide if it is pure using the above test results.

Carbon

There are three main **allotropes** of carbon:
- Diamond (see below)
- Graphite (see below)
- Buckminster fullerene (buckyball – see page 51).

Allotropes are different forms of the same element in the same physical state. The atoms of the element are arranged in a different molecular structure. The allotropes of carbon are all solids.

Diamond

Diamond has a rigid structure.
- It does not conduct electricity.
- It is **insoluble** in water.
- It is used in jewellery because it is colourless, clear (transparent) and lustrous (shiny).
- It can be used in cutting tools because it is very hard and has a very high melting point.

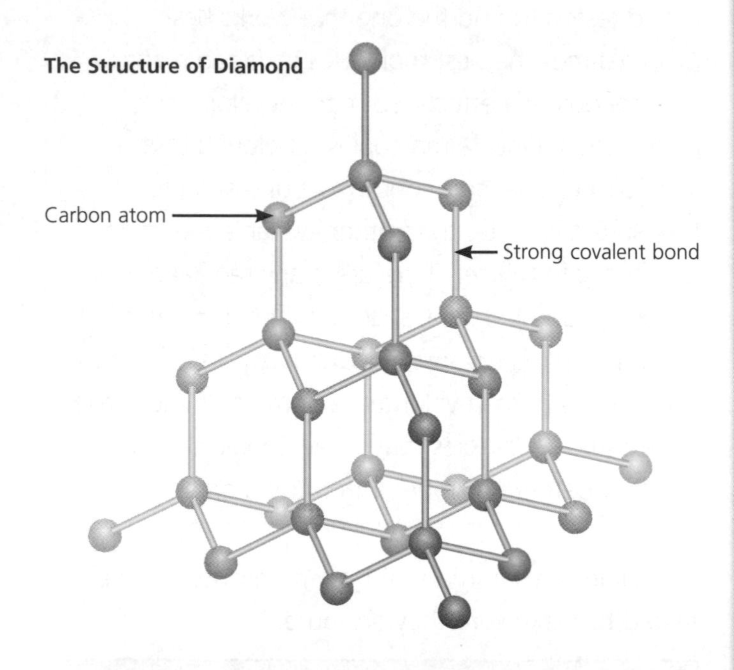

The Structure of Diamond

Graphite

Graphite has a layered structure.
- It is insoluble in water.
- It is black, which is why it is used in pencils.
- It is lustrous and opaque (light cannot travel through it).
- It conducts electricity and has a very high melting point, so is used to make electrodes for electrolysis.
- It is slippery, so it is used in lubricants.

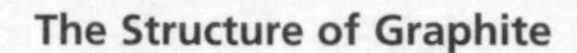
The Structure of Graphite

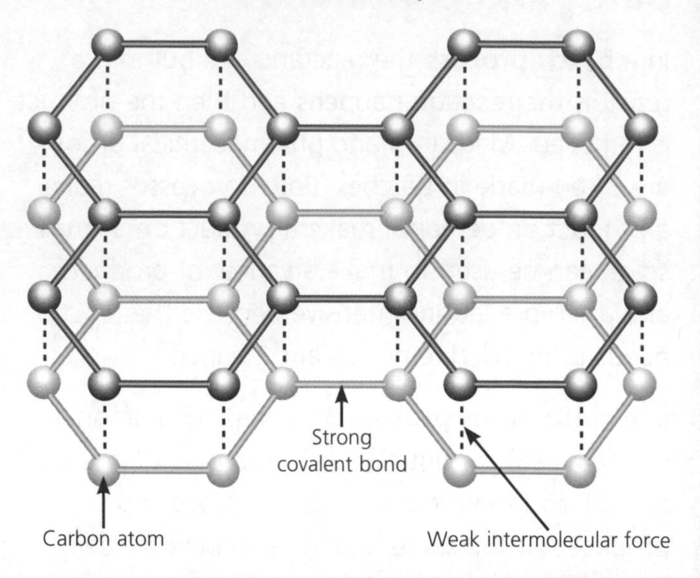

More on Diamond

Diamond is a giant molecule made of carbon atoms that are bonded to four other carbon atoms by strong covalent bonds. The giant structure has a large number of covalent bonds and this results in diamond having a high melting point because it needs lots of energy to break the bonds. It does not have any free electrons so it does not conduct electricity. Unlike graphite, it does not have separate layers and, because there are strong covalent bonds between the carbon atoms, it is very hard.

More on Graphite

Graphite is a giant molecule that exists in layers of carbon atoms bonded to three other carbon atoms by strong covalent bonds. The layers are held together by weak intermolecular forces, allowing each layer to slide easily, so graphite can be used as a lubricant. The presence of free (delocalised) electrons in graphite results in it being an electrical conductor. It also has a high melting point because the giant structure has many strong covalent bonds to break and lots of energy is needed to do this.

Diamond and graphite are giant molecular structures – many trillions of atoms joined together in a network by covalent bonds.

Nanochemistry

Nanochemistry deals with materials on an atomic scale (i.e. individual atoms), whereas chemistry is usually concerned with much larger quantities. Two forms of carbon that nanochemists are interested in are buckminster fullerene and nanotubes.

Buckminster Fullerene

Buckminster fullerene (C_{60}) consists of 60 carbon atoms arranged in a sphere.

Structure of Buckminster Fullerene

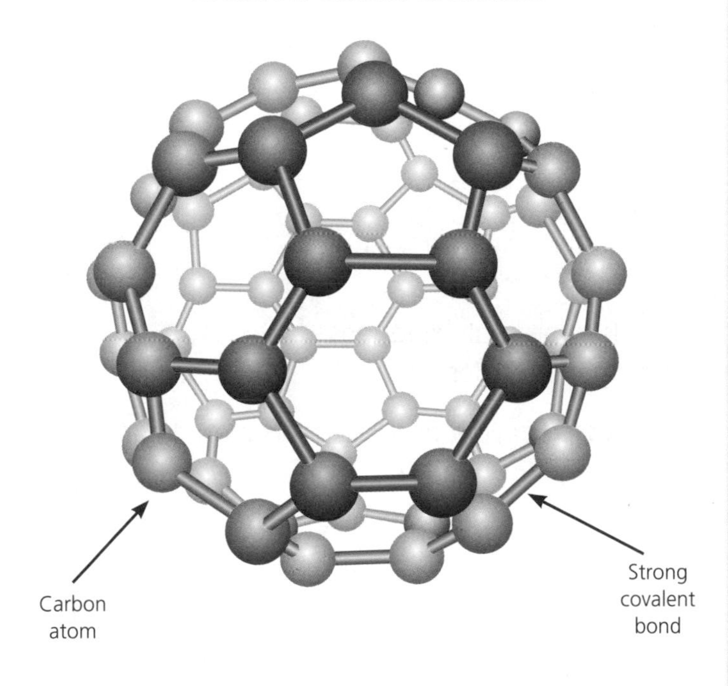

Nanotubes

The discovery of buckminster fullerene led chemists to investigate building similar structures.

In the early 1990s the first **nanotubes** were made by joining fullerenes together. They look like sheets of graphite hexagons curled over into a tube.

Nanotubes conduct electricity and are very strong. They are used to:

- reinforce graphite tennis rackets because of their strength
- make connectors and semiconductors in the most modern molecular computers because of their electrical properties
- develop new, more efficient industrial catalysts.

Structure of a Nanotube

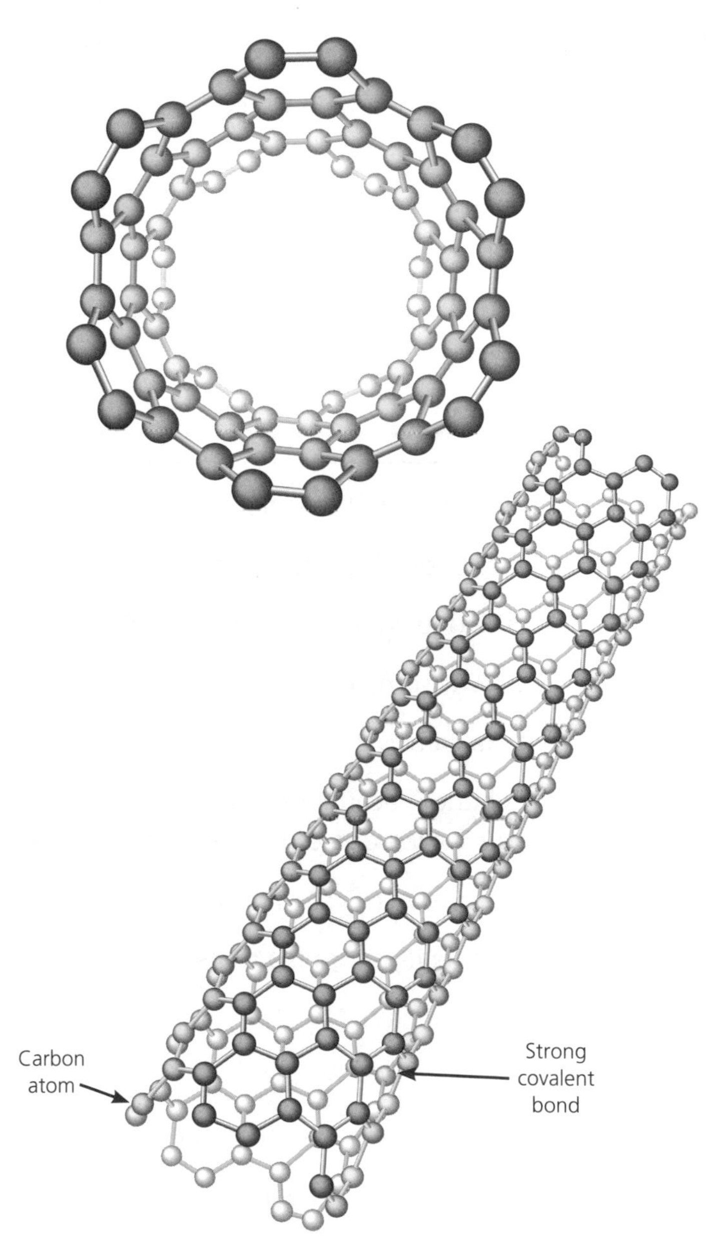

Fullerenes and nanotubes can be used to cage other molecules because they are the perfect shape to trap other substances inside them. The caged substances can be:

- **drugs**, e.g. a major new HIV treatment uses buckyballs to deliver a material which disrupts the way in which the HIV virus works

HT
- **catalysts** – by attaching catalyst material to a nanotube, a massive surface area can be achieved, making the catalyst very efficient.

C3 Exam Practice Questions

1 **a)** Explain what must happen to reactant particles for a reaction to take place. **[2]**

b) Describe what happens to particles when you increase the concentration. **[1]**

c) Look at the graph below. The curve from the original reaction is labelled.

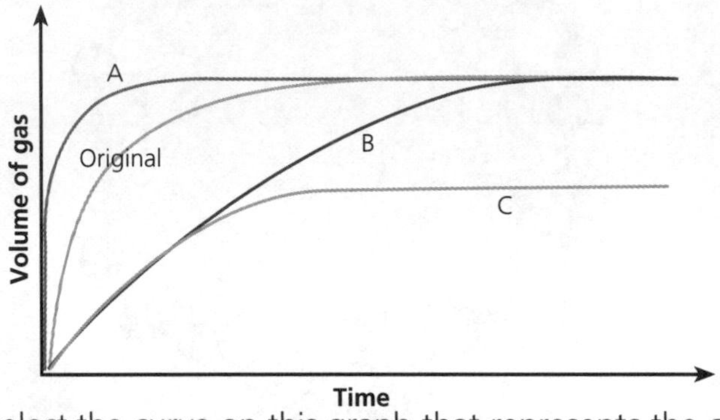

Volume of gas

A

Original

B

C

Time

Select the curve on this graph that represents the curve showing the same reaction but using:

i) a lower original temperature. **[1]**

ii) half the original amount of reactants. **[1]**

d) Explain why some collisions do not result in a successful reaction. **[2]**

2 **a)** What is a catalyst? **[2]**

b) Describe the precautions that have to be taken in flour mills to prevent explosions. Use your knowledge of rates of reaction to explain the danger of explosion. **[6]**

The quality of your written communication will be assessed in your answer to this question.

3 Calculate the relative formula mass (M_r) of:

a) sodium carbonate, Na_2CO_3 **[1]**

b) ethanol, C_2H_5OH **[1]**

c) In the following reaction, 12g of carbon reacts with 32g of oxygen: **carbon + oxygen → carbon dioxide**

i) Calculate the mass of carbon dioxide made in the reaction. **[1]**

ii) What mass of oxygen is needed to burn 1g of carbon? Give your answer to 2 significant figures. **[2]**

4 **a)** Diamond and graphite are both forms of which element? **[1]**

b) Explain why diamond is used in cutting tools. **[2]**

5 **a)** David heated 100g of water by burning 5g of petrol. The water temperature started at 21°C and went up to 47°C. Calculate the temperature change. **[1]**

HT **b)** Use David's results to calculate how much energy was made by burning the petrol. (Specific heat capacity of water = 4.2 J/g/°C) **[2]**

6 Use ideas about structure and bonding to explain why graphite can conduct electricity. **[3]**

7 **a)** Outline two factors that make researching a new drug expensive. **[2]**

b) Describe what actions must be taken before a drug can be sold in the UK. **[2]**

C4: The Periodic Table

This module looks at:
- The structure of the atom, protons, neutrons and electrons; isotopes.
- How metals and non-metals bond together, and the properties of ionic substances.
- Groups and periods, and properties of molecules.
- Properties of Group 1 metals, their reaction with water and flame tests.
- Properties of Group 7 non-metals, including their reaction with alkali metals.
- Properties of transition metals, decomposition of carbonates, and precipitation reactions.
- Metallic structure, properties of metals and superconductors.
- Sources and treatment of drinking water, pollution and testing for dissolved ions.

The Discovery of Atoms

The model of atomic structure has changed over time as new evidence has been discovered. The idea that everything is made from atoms was first put forward in ancient Greece but it was not until **John Dalton** published his 'Atomic Theory' in 1803 that the idea started to gain acceptance. He imagined atoms to look like billiard balls. His main ideas were:
- Elements are made from very tiny particles called atoms, which cannot be made or broken up.
- Atoms of the same element are identical and different to other elements' atoms.
- Atoms combine to make compounds.

We now know that the theory was not totally correct. **J. J. Thomson** discovered that even smaller particles could be pulled out of atoms using a very high voltage (the discovery of the electron). He pictured the atom to be solid like a currant bun and his experiments removed the currants! A few years later, **Ernest Rutherford** proved that the atom was not solid but had a very dense centre called a nucleus. **Niels Bohr** combined his work on the orbits of electrons to develop the idea of the solar system atom, the nucleus in the centre like the Sun with the electrons orbiting around it like planets.

The ideas at each stage did not fully explain the behaviour of atoms. All the evidence had to be tested and predictions had to be made that would all have to be correct before a new theory was accepted.

HT Each development was a step closer to understanding the atom and helped other scientists with their work. Not all experiments went to plan: Rutherford's assistants, Geiger and Marsden, expected the α particles they were firing at gold foil to punch through in a straight line. A few were deflected and some even bounced straight back leading to the idea that the α particles were passing very close to, or hitting, the nucleus of the gold atoms.

Structure of an Atom

All substances are made up of very small particles called **atoms** with a very low mass. Atoms have a central **nucleus**. The nucleus is positively charged because it is made up of **protons** which are positively charged and **neutrons** which are neutral. The space around the nucleus is occupied by negatively charged **electrons** arranged in shells. Overall, an atom has **no charge**.

HT An atom has the same number of protons and electrons, so they cancel out each other's charges. The mass of an atom is about 10^{-23}g and it has a radius of 10^{-10}m.

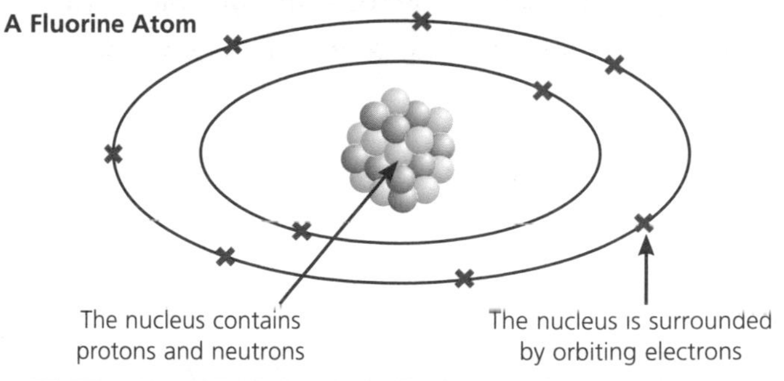

Atomic Particle	Relative Charge	Relative Mass
Proton	+1	1
Neutron	0	1
Electron	–1	0.0005 (zero)

C4 Atomic Structure

Elements and Compounds

An **element** is a substance that cannot be broken down chemically and contains only one type of atom.

A **compound** is a substance that contains at least two elements that are **chemically combined**. You can find out which elements make up a compound by looking at the compound's **formula** and identifying the elements from the Periodic Table, for example:

- sodium chloride (NaCl) contains the elements sodium (Na) and chlorine (Cl)
- potassium nitrate (KNO_3) contains the elements potassium (K), nitrogen (N) and oxygen (O).

Atomic Number

The numbers next to the elements in the Periodic Table give information about the element.

The elements in the Periodic Table are arranged in order of ascending atomic number, starting with hydrogen (atomic number 1) at the top. The **atomic number** is the number of protons in the atom.

You can use the Periodic Table to find an element if you know its atomic number. Likewise, if you know where the element is in the Periodic Table you can find its atomic number.

Isotopes

All atoms of a particular element have the same number of protons. (Different numbers of protons indicate atoms of different elements.) However, some elements have varieties that have different numbers of neutrons; these are called **isotopes**.

Isotopes have the same atomic number but a different mass number. The **mass number** is the total number of protons and neutrons in an atom.

Mass number = Number of protons + Number of neutrons

More on Atomic Number and Isotopes

HT

The table below shows some common elements and how the number of protons, neutrons and electrons can be calculated from the atomic number and mass number.

Chemical symbol	Chemical name	Number of protons	Number of electrons	Number of neutrons
$^{1}_{1}\mathbf{H}$	Hydrogen	1	1	0 (1 – 1 = 0)
$^{3}_{1}\mathbf{H}$	Hydrogen	1	1	2 (3 – 1 = 2)
$^{4}_{2}\mathbf{He}$	Helium	2	2	2 (4 – 2 = 2)
$^{16}_{8}\mathbf{O}$	Oxygen	8	8	8 (16 – 8 = 8)
$^{23}_{11}\mathbf{Na}$	Sodium	11	11	12 (23 – 11 = 12)

$^{3}_{1}\mathbf{H}$ is also called the hydrogen-3 isotope

Electronic Structure

The electronic configuration tells us how the electrons are arranged around the nucleus.

Aluminium has the electronic structure 2.8.3. This means it has 3 occupied shells and a total of 13 electrons. The element with an electronic structure of 2.6 has 8 electrons; this will be the same as its atomic number and we can see that this is oxygen by looking it up on the Periodic Table.

- The electrons in an atom occupy the lowest available shells.
- The first energy level or shell can contain a maximum of only 2 electrons.
- The shells after this can each hold a maximum of 8 electrons – an outer shell containing 8 electrons is known as a full outer shell.
- The electronic structure of an ion can be determined by working out the electronic structure of the atom and adding electrons for a negative ion or subtracting electrons for a positive ion.

The Modern Periodic Table – Electronic Structure of the First 20 Elements

Hydrogen, H
Atomic No. = 1
No. of electrons = 1
1

GROUP 0

Helium, He
Atomic No. = 2
No. of electrons = 2
2

GROUP 1

Lithium, Li
Atomic No. = 3
No. of electrons = 3
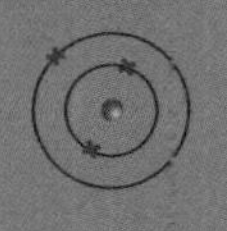
2.1

Sodium, Na
Atomic No. = 11
No. of electrons = 11
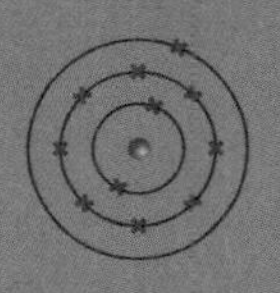
2.8.1

Potassium, K
Atomic No. = 19
No. of electrons = 19
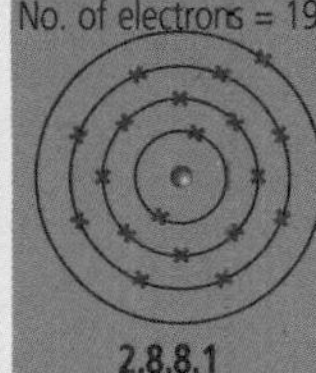
2.8.8.1

GROUP 2

Beryllium, Be
Atomic No. = 4
No. of electrons = 4
2.2

Magnesium, Mg
Atomic No. = 12
No. of electrons = 12
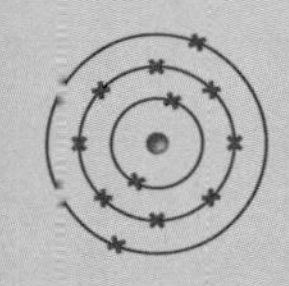
2.8.2

Calcium, Ca
Atomic No. = 20
No. of electrons = 20
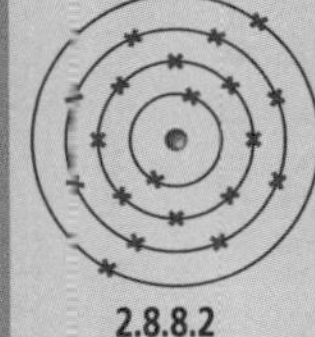
2.8.8.2

GROUP 3

Boron, B
Atomic No. = 5
No. of electrons = 5
2.3

Aluminium, Al
Atomic No. = 13
No. of electrons = 13
2.8.3

GROUP 4

Carbon, C
Atomic No. = 6
No. of electrons = 6
2.4

Silicon, Si
Atomic No. = 14
No. of electrons = 14
2.8.4

GROUP 5

Nitrogen, N
Atomic No. = 7
No. of electrons = 7
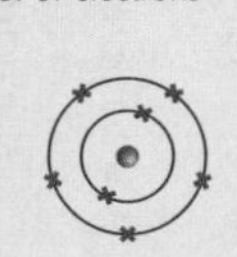
2.5

Phosphorus, P
Atomic No. = 15
No. of electrons = 15
2.8.5

GROUP 6

Oxygen, O
Atomic No. = 8
No. of electrons = 8
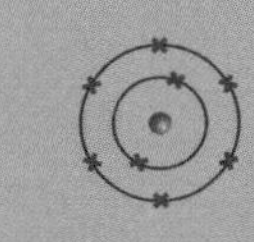
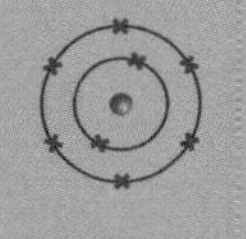
2.6

Sulfur, S
Atomic No. = 16
No. of electrons = 16
2.8.6

GROUP 7

Fluorine, F
Atomic No. = 9
No. of electrons = 9
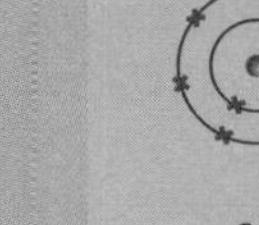
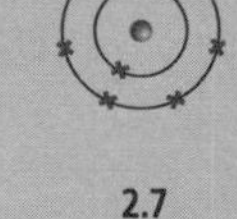
2.7

Chlorine, Cl
Atomic No. = 17
No. of electrons = 17
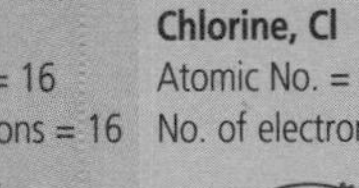
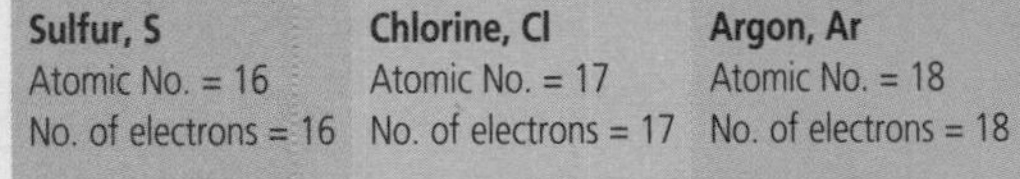
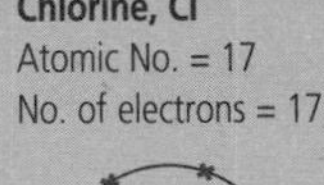
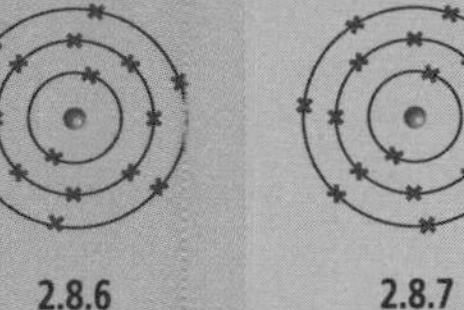
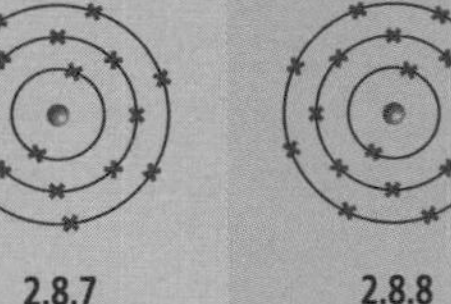
2.8.7

Neon, Ne
Atomic No. = 10
No. of electrons = 10
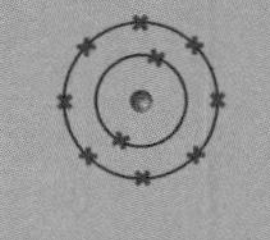
2.8

Argon, Ar
Atomic No. = 18
No. of electrons = 18
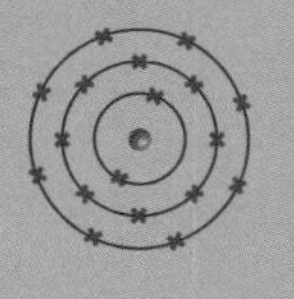
2.8.8

This table is arranged in order of atomic (proton) number, placing the elements in groups.

Elements in the same group have the same number of electrons in their highest occupied energy level (outer shell).

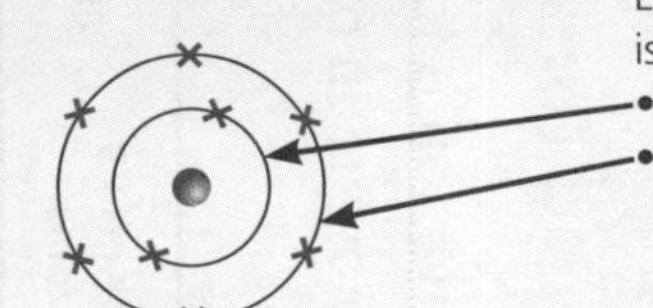

Electron configuration of oxygen is 2.6 because there are:
- 2 electrons in the first shell
- 6 electrons in the second shell.

C4 Ionic Bonding

Ions, Atoms and Molecules

An **uncharged** particle is either:

- an atom on its own, e.g. Na, Cl
- a molecule with two or more atoms bonded together, e.g. Cl_2, CO_2.

An **ion** is a charged atom or group of atoms, e.g. Na^+, Cl^-, NH_4^+, SO_4^{2-}.

Forming Ions

A **positive ion** is formed when an atom (usually a metal) or group of atoms loses one or more electrons.

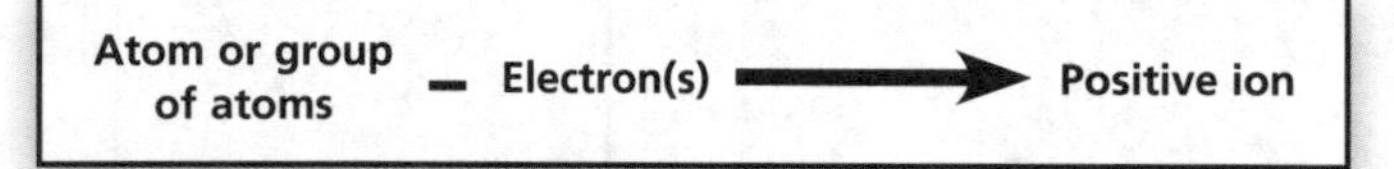

- Losing 1 electron makes a + ion, e.g. Na^+.
- Losing 2 electrons makes a 2+ ion, e.g. Mg^{2+}.

A **negative ion** is formed when an atom (usually a non-metal) or group of atoms gains one or more electrons.

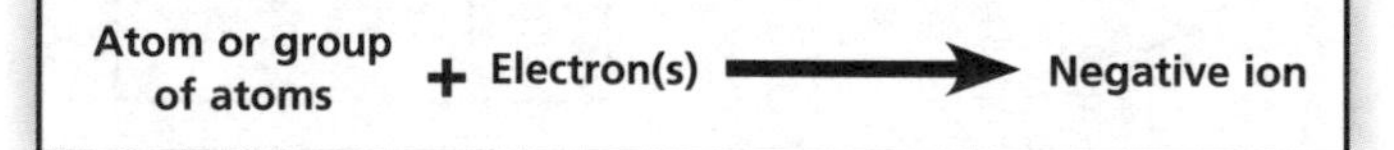

- Gaining 1 electron makes a – ion, e.g. Cl^-.
- Gaining 2 electrons makes a 2– ion, e.g. O^{2-}.

Atoms form ions to gain a full outer shell of electrons; this is a very stable structure.

Ionic Bonding

A **metal** and a **non-metal** combine by **transferring electrons**. The metal atoms transfer electrons to become positive ions and the non-metal atoms receive electrons to become negative ions.

The positive ions and negative ions are then **attracted** to each other. Two compounds that are **bonded** by the attraction of oppositely charged ions are sodium chloride and magnesium oxide.

Sodium chloride has a high melting point. It does not conduct electricity when it is solid. However, it can dissolve in water, and the solution produced can conduct electricity. It is also able to conduct electricity when it is molten.

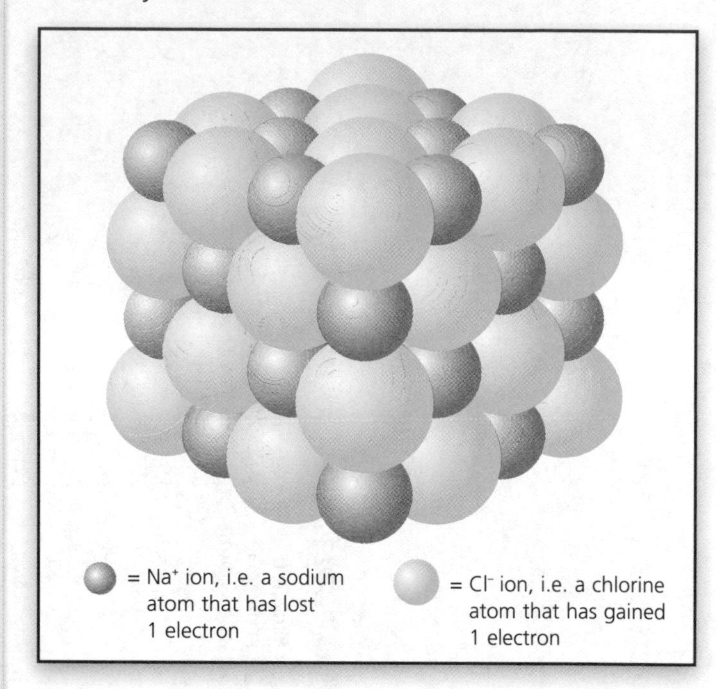

Magnesium oxide has an even higher melting point. It does not conduct electricity when it is solid. However, it can conduct electricity when it is molten.

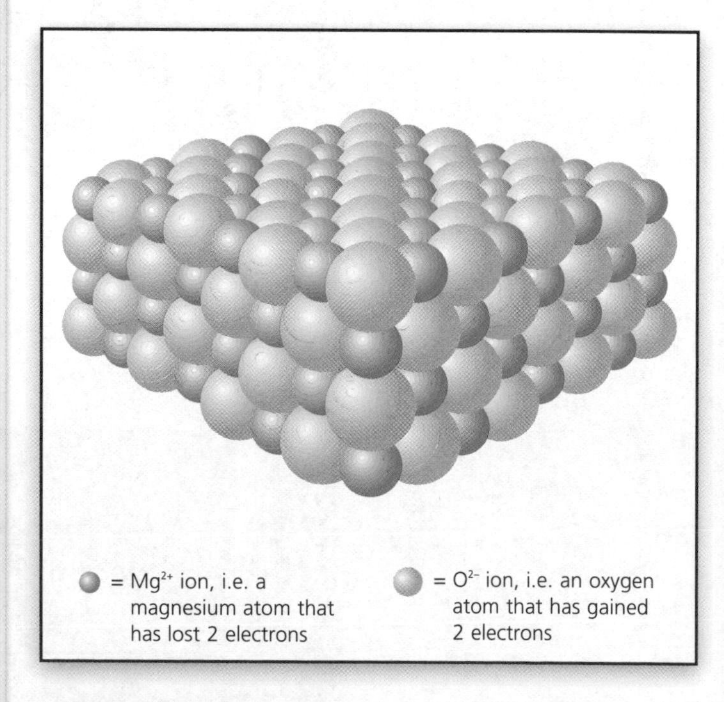

Sodium chloride (NaCl) and magnesium oxide (MgO) form **giant ionic lattices** in which positive ions and negative ions are **electrostatically** attracted to each other.

Structure and Physical Properties of Sodium Chloride and Magnesium Oxide

The strong attraction between oppositely charged ions has to be overcome for them to melt and this results in sodium chloride and magnesium oxide having high melting points.

The attraction is stronger in magnesium oxide than in sodium chloride and so magnesium oxide has a higher melting point.

They conduct electricity when molten or in solution because the charged ions are free to move about.

When they are solid, the ions are held in place and cannot move about, which means they do not conduct electricity.

Formulae of Ionic Compounds

All ionic compounds are neutral substances that have equal charges on the positive ion(s) and negative ion(s).

The table below shows how ions with different charges combine to form ionic compounds.

The Ionic Bond

When a **metal** and a **non-metal** combine, electrons are transferred from one atom to the other to form **ions**. Each ion then has a **complete outer shell** (a **stable octet**).

Example 1

The sodium atom has 1 electron in its outer shell which is transferred to the chlorine atom to give them both 8 electrons in their outer shell.

The atoms become ions (Na^+ and Cl^-) and the compound formed is sodium chloride, NaCl.

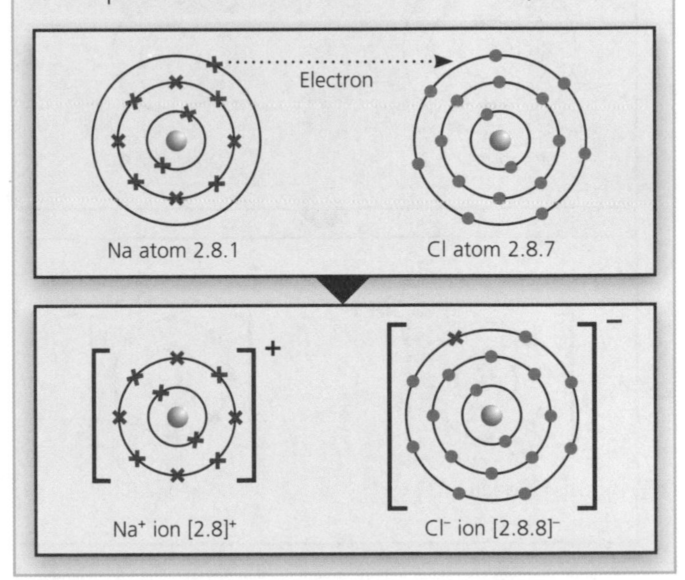

		Negative Ions			
		1– e.g. Cl^-, OH^-		2– e.g. SO_4^{2-}, O^{2-}	
Positive Ions	1+ e.g. K^+, Na^+	KCl (1+, 1–)	NaOH (1+, 1–)	K_2SO_4 (2 × 1+ = 2+, 2–)	Na_2O (2 × 1+ = 2+, 2–)
	2+ e.g. Mg^{2+}, Cu^{2+}	$MgCl_2$ (2+, 2 × 1– = 2–)	$Cu(OH)_2$ (2+, 2 × 1– = 2–)	$MgSO_4$ (2+, 2–)	CuO (2+, 2–)
	3+ e.g. Al^{3+}, Fe^{3+}	$AlCl_3$ (3+, 3 × 1– = 3–)	$Fe(OH)_3$ (3+, 3 × 1– = 3–)	$Al_2(SO_4)_3$ (2 × 3+ = 6+, 3 × 2– = 6–)	Fe_2O_3 (2 × 3+ = 6+, 3 × 2– = 6–)

HT The Ionic Bond (cont)

Example 2

The magnesium atom has 2 electrons in its outer shell which are transferred to the oxygen atom to give them both 8 electrons in their outer shell.

The atoms become ions (Mg^{2+} and O^{2-}) and the compound formed is magnesium oxide, MgO.

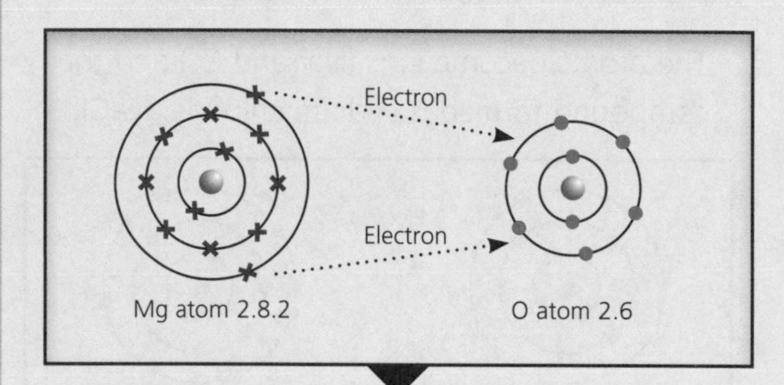

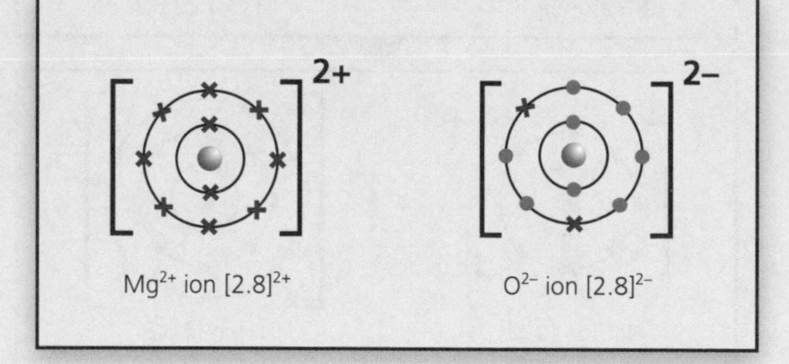

Example 3

magnesium + chlorine → magnesium chloride

The magnesium atom has 2 electrons in its outer shell. A chlorine atom only needs 1 electron, therefore, 2 chlorine atoms are needed.

The atoms become ions (Mg^{2+}, Cl^- and Cl^-) and the compound formed is magnesium chloride, $MgCl_2$.

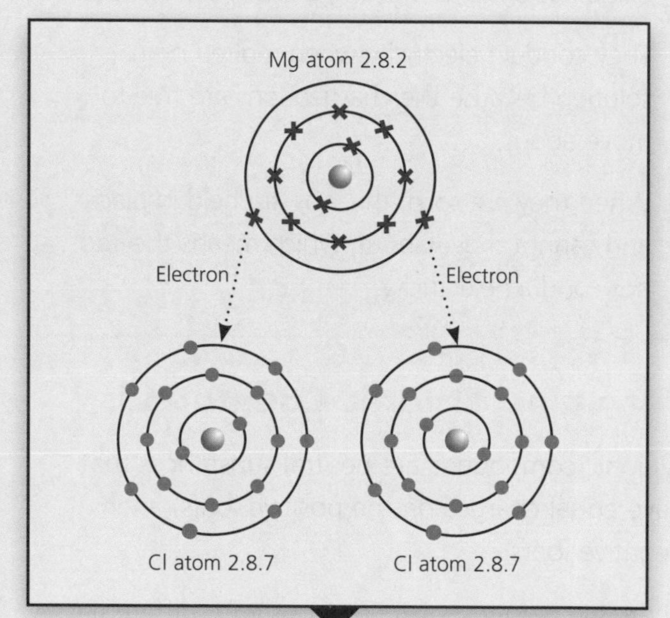

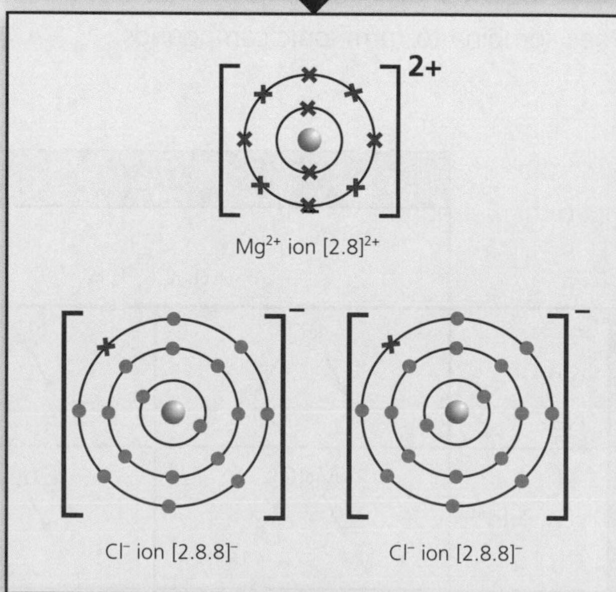

The Periodic Table

Elements are the building blocks of all materials. The 100 or so elements are arranged in order of **ascending atomic number**, and then arranged in rows (periods) so that elements with similar properties are in the same column (group). This forms the basis of the **Periodic Table**. More than three-quarters of the elements are metals. The others are non-metals.

The Periodic Table is more than just a list of elements. It has a pattern that is linked to the elements' properties.

Johann Döbereiner was the first scientist to try to arrange the elements in a pattern. He linked elements that had similar properties and found that they made **triads**, i.e. groups of three elements, e.g. chlorine, bromine and iodine. He noticed that the atomic weight of the triad's second element was the average of the other two. Although some elements could not be put into a triad, he published his idea in 1817.

In 1864 **John Newlands** arranged the elements by increasing atomic weight. He noticed that every eighth element had similar properties and coincided with the triads. Other scientists did not accept his **Law of Octaves** because some elements had not yet been discovered and some did not fit the pattern.

Dimitri Mendeleev refined Newlands' ideas by putting all the known data for each element onto cards which he arranged in groups and periods in weight order based on their properties. If the weight order contradicted the properties of two elements he swapped them around to maintain his patterns. He left gaps for undiscovered elements and predicted their properties.

Mendeleev's ideas were published in 1869 but were not accepted. A few years later, a missing element was discovered. The properties matched his predictions and the Periodic Table was established.

HT Mendeleev's periodic table became widely accepted after the discovery of gallium, which had the properties he had predicted. The later discovery of electrons and the structure of atoms helped to explain Mendeleev's patterns. The number of electrons in each shell matched the rows in his periodic table.

Mendeleev had ignored atomic weight order for some elements and swapped them to match their properties. He was shown to be correct when the elements were arranged in atomic number order as in the modern Periodic Table.

The atomic weight order was incorrect due to the existence of heavier isotopes of elements with the lower atomic number (e.g. tellurium – atomic number 52, A_r 128; iodine – atomic number 53, A_r 127).

Groups and Periods

A vertical column of elements in the Periodic Table is called a **group**, e.g. Group 1 contains lithium (Li), sodium (Na) and potassium (K), among others. Elements in the same group have similar chemical properties. This is because they have the **same number of electrons in their outer shell**. This number also coincides with the group number, e.g. Group 1 elements have 1 electron in their outer shell and Group 7 elements have 7 electrons in their outer shell.

A **horizontal row** of elements in the Periodic Table is called a **period**, e.g. lithium (Li), carbon (C) and neon (Ne) are all elements in the second period.

The period to which the element belongs corresponds to the **number of shells of electrons** it has, e.g. sodium (Na), aluminium (Al) and chlorine (Cl) all have three shells of electrons so they are found in the third period.

Bonding

A **molecule** is two or more atoms bonded together. There are two types of bonding:

- **Covalent bonding** – non-metals combine by sharing electrons.
- **Ionic bonding** – metals and non-metals combine by transferring electrons.

Two examples of covalently bonded molecules are **water** and **carbon dioxide**. Water, H_2O, is a liquid with a low melting point, which does not conduct electricity.

Carbon dioxide, CO_2, is a gas with a low melting point, which does not conduct electricity.

Water and carbon dioxide are simple molecules. They have low melting points because the forces between the molecules (intermolecular forces) are weak, so only a small amount of energy is needed to overcome them.

HT Representing Molecules

The following are all examples of **covalently bonded molecules**. You need to be familiar with how they are formed:

Water (H_2O) – the outer shells of the hydrogen and oxygen atoms overlap, and the oxygen atom shares a pair of electrons with each hydrogen atom to form a water molecule:

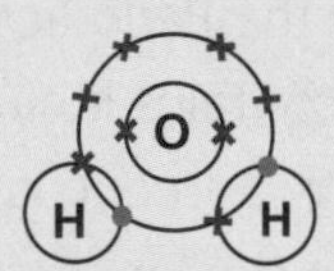

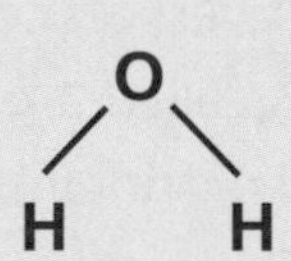

Hydrogen (H_2) – the two hydrogen atoms share a pair of electrons:

H – H

Methane (CH_4) – the carbon atom shares a pair of electrons with each hydrogen atom:

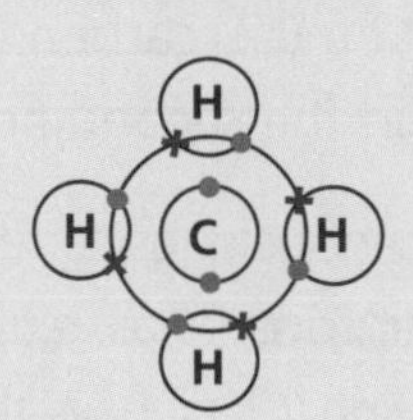

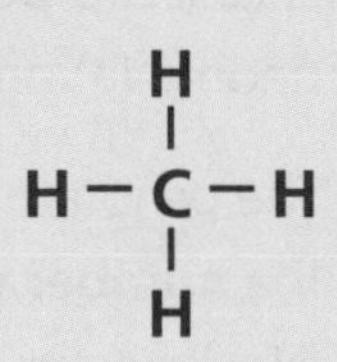

Chlorine (Cl_2) – the two chlorine atoms share a pair of electrons:

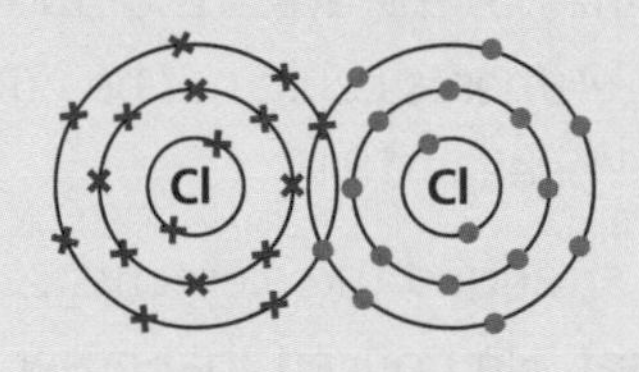

Cl – Cl

Carbon dioxide (CO_2) – the outer shells of the carbon and oxygen atoms overlap and the carbon atom shares two pairs of electrons with each oxygen atom to form a double covalently bonded molecule:

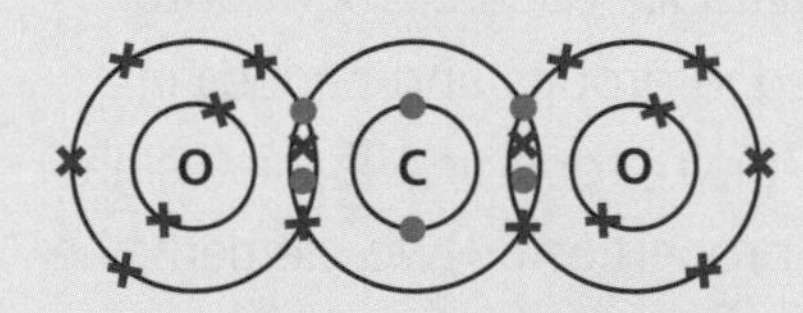

O = C = O

Properties of Simple Covalently Bonded Molecules

The bond between two atoms in a simple covalently bonded molecule (e.g. water or carbon dioxide) is very strong.

The **intermolecular forces** of attraction *between* molecules are weak. This results in them having low melting points. They do not conduct electricity as they do not have any free electrons.

Group 1 – The Alkali Metals

The **alkali metals** occupy the first vertical column (**Group 1**) at the left-hand side of the Periodic Table. The first three elements in the group are lithium, sodium and potassium. They all have one electron in their outer shell which means they have similar properties. Alkali metals are stored under oil because they react with air and react vigorously with water.

Flame Tests

Lithium, sodium and potassium compounds can be recognised by the colours they produce in a **flame test**. The method used is explained below:

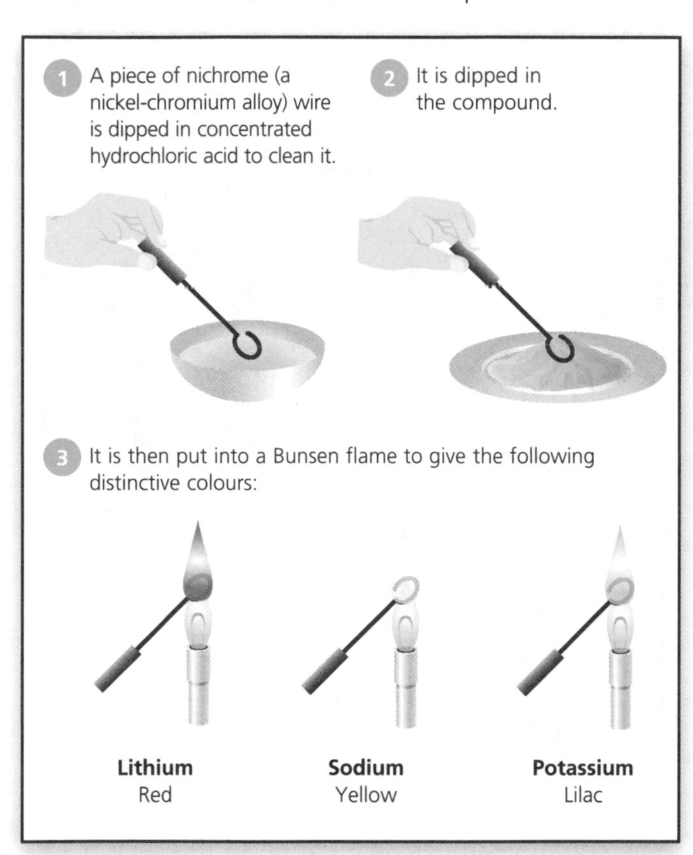

Reacting Alkali Metals with Water

Alkali metals react with water to produce hydrogen and a **hydroxide.** Alkali metal hydroxides are soluble and form **alkaline solutions**.

As we go **down** the group, the alkali metals become more reactive and so they react more vigorously with water. They all float and some may melt and produce hydrogen gas (which may ignite).

Lithium reacts gently with water, sodium reacts more aggressively and potassium reacts so aggressively that it melts and burns with a lilac flame.

The diagram below shows what happens when a small piece of potassium is dropped into water:

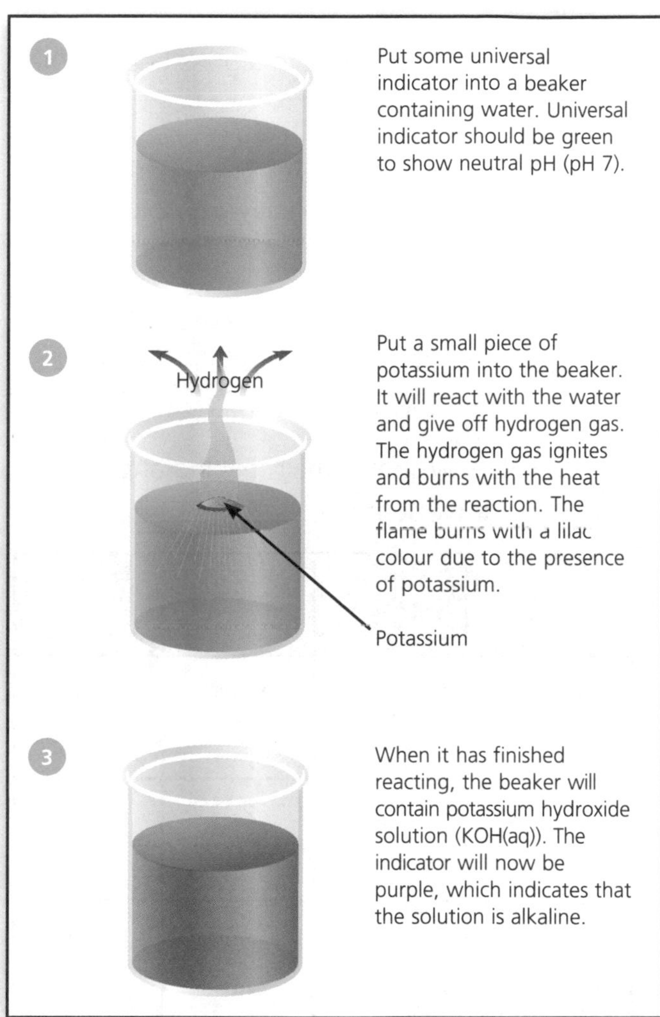

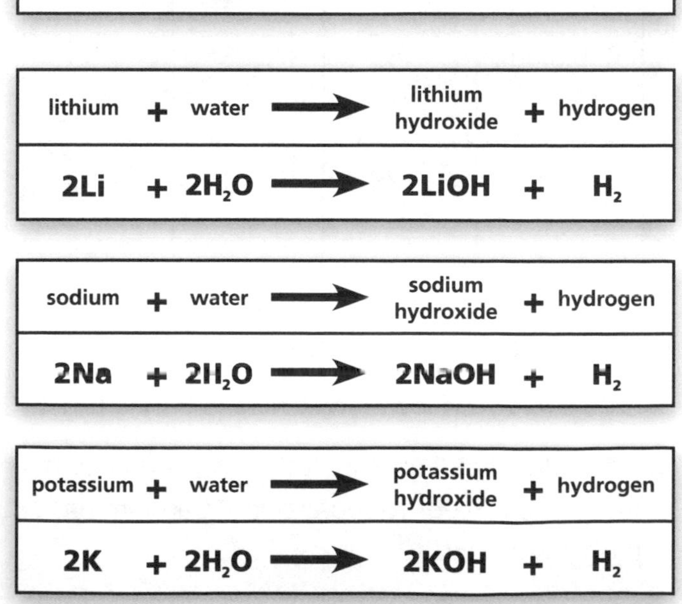

lithium	+	water	⟶	lithium hydroxide	+	hydrogen
$2Li$	+	$2H_2O$	⟶	$2LiOH$	+	H_2

sodium	+	water	⟶	sodium hydroxide	+	hydrogen
$2Na$	+	$2H_2O$	⟶	$2NaOH$	+	H_2

potassium	+	water	⟶	potassium hydroxide	+	hydrogen
$2K$	+	$2H_2O$	⟶	$2KOH$	+	H_2

C4 The Group 1 Elements

Properties of the Alkali Metals

Rubidium is the fourth element in Group 1. Rubidium, like the other elements in the group, reacts with water. The reaction would be:

- very fast
- exothermic
- violent (if it is carried out in a glass beaker, the beaker may shatter).

Caesium (the fifth element in the group) is even more reactive and will react very violently with water.

HT Even though the alkali metals have similar chemical properties, their physical properties alter as we go down the group.

The table below shows their melting and boiling points and their densities:

Element	Melting Point (°C)	Boiling Point (°C)	Density (g/cm^3)
Lithium, Li	180	1340	0.53
Sodium, Na	98	883	0.97
Potassium, K	64	760	0.86
Rubidium, Rb	39	688	1.53
Caesium, Cs	29	671	1.90

The melting points and boiling points of alkali metals decrease going down the group.

Caesium has the lowest melting and boiling points.

Generally, the density increases as we go down the group (except for potassium). Caesium has the largest density.

Trends in Group 1

Alkali metals have similar properties because when they react, an atom loses one electron to form a **positive ion** with a **stable electronic structure**, i.e. it has a full outer shell of electrons.

The alkali metals become more reactive as we go down the group because the outer shell gets further away from the influence of the nucleus, making it easier for an atom to lose an electron from this shell.

Oxidation involves the loss of electrons from an atom. Examples are shown below:

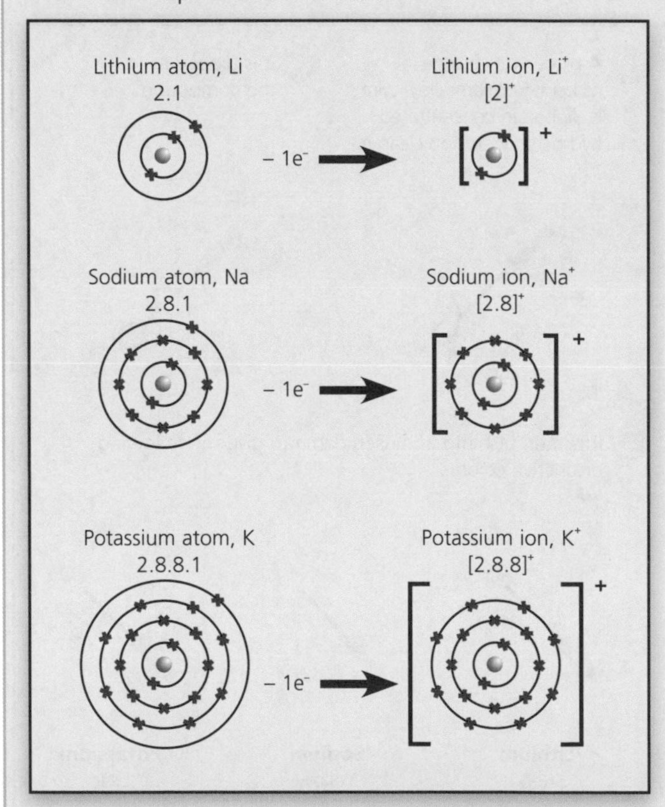

The equations for the formation of the Group 1 metal ions are usually written as follows:

$$Li - e^- \longrightarrow Li^+$$

$$Na - e^- \longrightarrow Na^+$$

$$K - e^- \longrightarrow K^+$$

Group 7 – The Halogens

There are five non-metals in Group 7 and they are known as the **halogens**. They all have seven electrons in their outer shell which means that they have similar chemical properties. You need to know about four of the halogens: **fluorine**, **chlorine**, **bromine** and **iodine**.

At room temperature:

- chlorine is a green gas
- bromine is an orange liquid
- iodine is a grey solid.

Iodine is used as an antiseptic to sterilise wounds.

Chlorine, the most commonly used halogen, is used to sterilise water and to make pesticides and plastics.

Reactions with Alkali Metals

Halogens react vigorously with alkali metals to form metal halides.

Use these examples to work out the product and write the word equation for any combination of halogen and alkali metal:

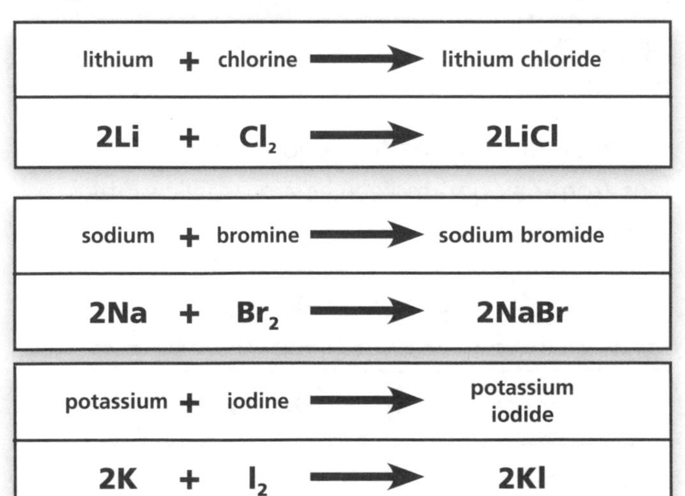

lithium + chlorine ⟶ lithium chloride

$2Li + Cl_2 \longrightarrow 2LiCl$

sodium + bromine ⟶ sodium bromide

$2Na + Br_2 \longrightarrow 2NaBr$

potassium + iodine ⟶ potassium iodide

$2K + I_2 \longrightarrow 2KI$

Displacement Reactions

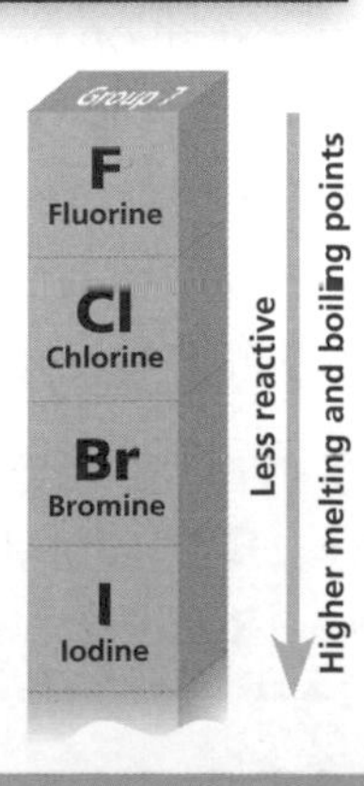

As we go down the group, the halogens become **less reactive**, and their melting and boiling points increase.

Fluorine is therefore the most reactive halogen and iodine is the least reactive.

A more reactive halogen will displace a less reactive halogen from an aqueous solution of its metal **halide**, i.e. chlorine will displace bromides and iodides, and bromine will displace iodides.

If chlorine gas is passed through an aqueous solution of potassium bromide, bromine is formed due to the displacement reaction taking place:

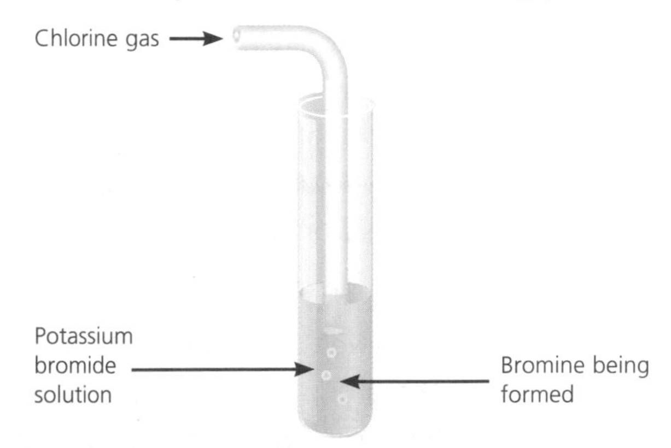

The table below shows the products of the reactions between halogens and aqueous solutions of halide salts:

	Potassium chloride	Potassium bromide	Potassium iodide
Chlorine Cl_2	X	Potassium chloride + bromine	Potassium chloride + iodine
Bromine Br_2	No reaction	X	Potassium bromide + iodine
Iodine I_2	No reaction	No reaction	X

X = No experiment conducted.

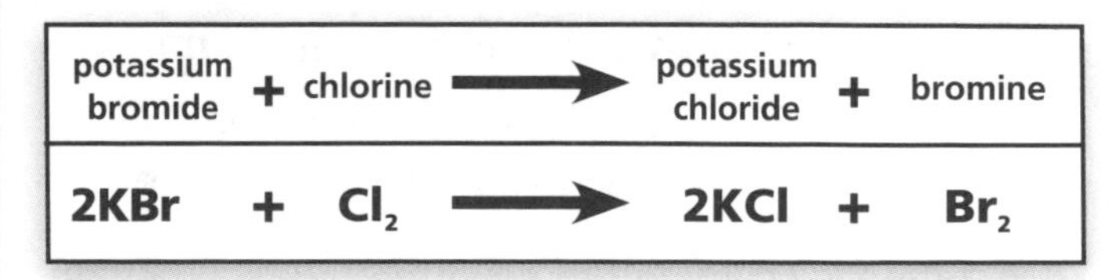

potassium bromide + chlorine ⟶ potassium chloride + bromine

$2KBr + Cl_2 \longrightarrow 2KCl + Br_2$

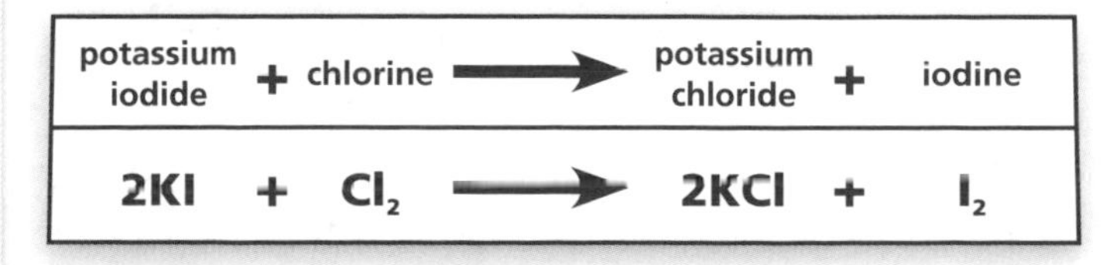

potassium iodide + chlorine ⟶ potassium chloride + iodine

$2KI + Cl_2 \longrightarrow 2KCl + I_2$

potassium iodide + bromine ⟶ potassium bromide + iodine

$2KI + Br_2 \longrightarrow 2KBr + I_2$

Properties of the Halogens

Fluorine is the first element in Group 7, and is the most reactive element in the group. It will displace all of the other halogens from an aqueous solution of their metal halides.

Astatine is the fifth element in Group 7. It is a semi-metallic, radioactive element and only very small amounts of it exist naturally. Even so, it is the least reactive of the halogens and, theoretically, it would be unable to displace any of the other halogens from an aqueous solution of their metal halides.

The **physical properties** of the halogens alter as we go down the group. The table below shows their melting and boiling points and their densities:

Element	Melting Point (°C)	Boiling Point (°C)	Density (g/cm^3)
Fluorine	–220	–188	0.0016
Chlorine	–101	–34	0.003
Bromine	–7	59	3.12
Iodine	114	184	4.95
Astatine	302 (estimated)	337 (estimated)	about 7 (estimated)

The melting points and boiling points of the halogens increase going down the group. The density also increases going down the group.

Astatine is estimated to have the highest melting point, boiling point and density. The figure for density in the table is estimated by looking at the trend of differences in density between each member of the group and adding that difference to the value for iodine. Accurate figures for astatine are not known because astatine is a very unstable element.

Trends in Group 7

The halogens have similar properties because when they react, an atom **gains** one electron to form a **negative ion** with a stable electronic structure, i.e. it has a full outer shell of electrons. **Reduction** involves the gain of electrons by an atom. Examples are shown below:

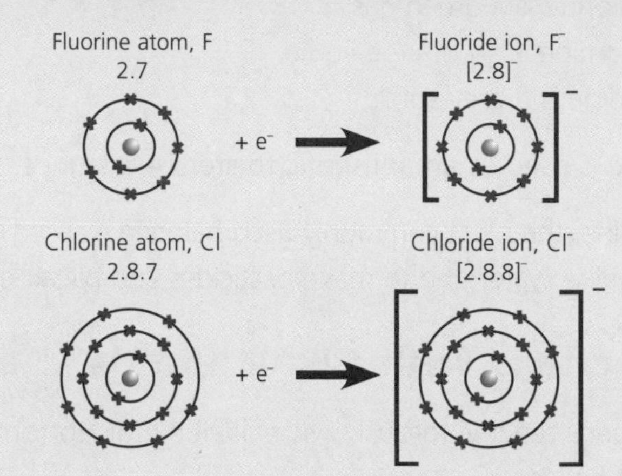

Fluorine is the most reactive because it is easiest for it to gain the extra electron.

The halogens become less reactive as we go down the group because the outer shell gets further away from the influence of the nucleus, making it harder for an atom to gain an electron.

The equations for the formation of the halide ions from halogen molecules are usually written as follows:

$$F_2 + 2e^- \rightarrow 2F^-$$

$$Cl_2 + 2e^- \rightarrow 2Cl^-$$

You can decide whether a reaction is an example of oxidation or reduction by looking at its equation, such as the ones above. If electrons are added, then it is a reduction reaction, and if electrons are taken away it is an oxidation reaction.

An easy way to remember the definitions of oxidation and reduction is **OIL RIG:**
- **O**xidation **I**s **L**oss of electrons
- **R**eduction **I**s **G**ain of electrons.

The Transition Metals

In the centre of the Periodic Table, between Groups 2 and 3, is a block of metallic elements called the **transition metals**. This block includes **iron** (Fe), **copper** (Cu), **platinum** (Pt), **mercury** (Hg), **chromium** (Cr) and **zinc** (Zn).

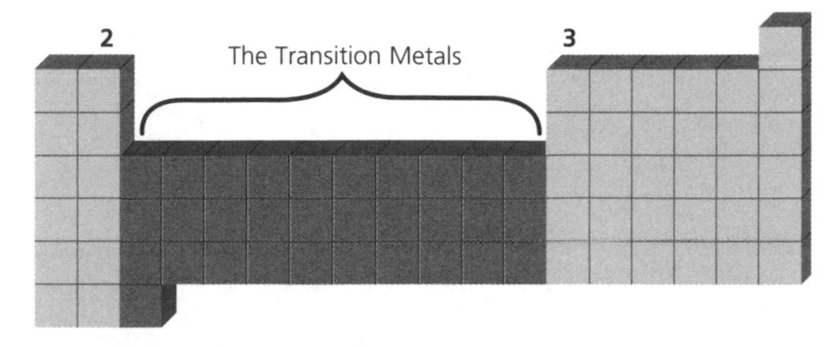

Transition metals have the typical properties of metals.

Compounds of transition metals are often **coloured**:

- copper compounds are blue
- iron(II) compounds are grey–green
- iron(III) compounds are orange–brown.

Many transition metals and their compounds can be used as **catalysts** in chemical reactions, for example:

- iron is used in the Haber process
- nickel is used in the manufacture of margarine.

Thermal Decomposition of Transition Metal Carbonates

Thermal decomposition is a reaction in which a substance is broken down into simpler substances by heating. When transition metal carbonates are heated, a **colour change** occurs and they decompose to form a **metal oxide** and **carbon dioxide**. For example, if copper carbonate is heated, the blue–green copper carbonate decomposes into black copper oxide and carbon dioxide, which turns limewater milky:

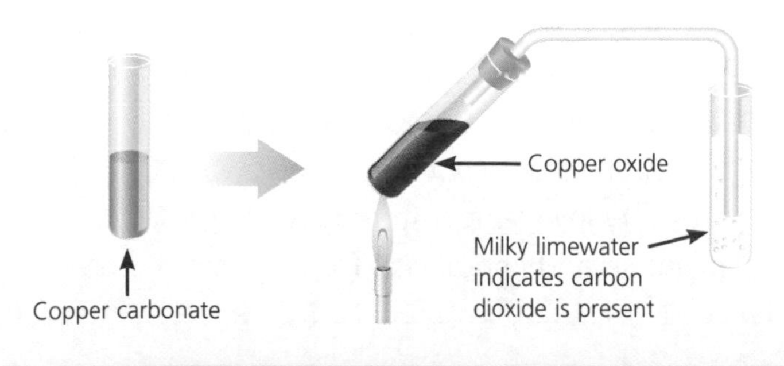

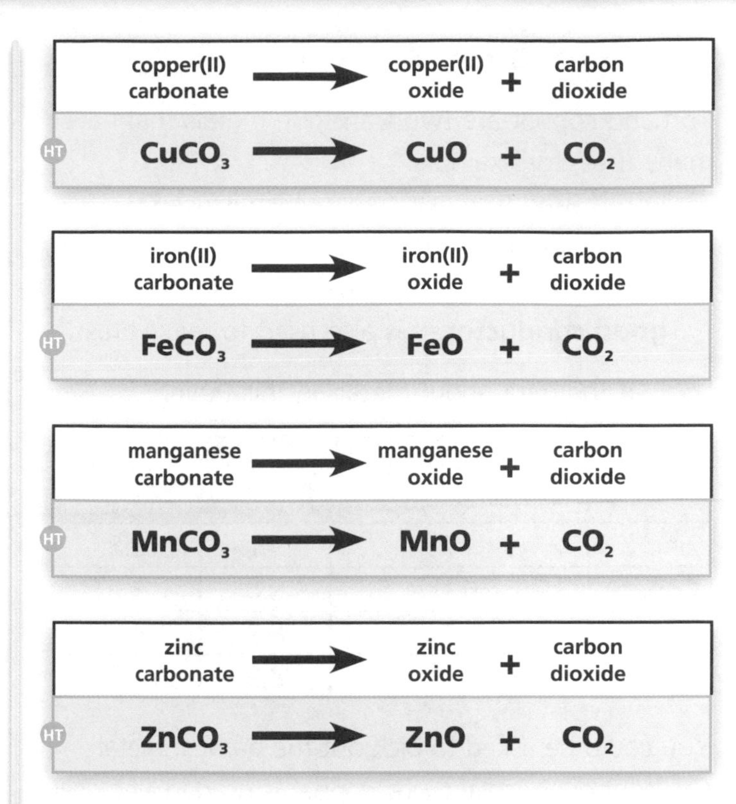

Identifying Transition Metal Ions

Metal compounds in solution contain metal ions. Some of these form **precipitates** (insoluble solids) that come out of solution when sodium hydroxide solution is added to them. The following ions form coloured precipitates:

Metal Ion	Colour of Precipitate	Ionic Symbol Equation
Copper(II), Cu^{2+}	Blue	HT $Cu^{2+} + 2OH^- \rightarrow Cu(OH)_2$
Iron(II), Fe^{2+}	Grey–green	HT $Fe^{2+} + 2OH^- \rightarrow Fe(OH)_2$
Iron(III), Fe^{3+}	Orange–brown	HT $Fe^{3+} + 3OH^- \rightarrow Fe(OH)_3$

The reaction between the transition metal ions and sodium hydroxide solution is called a **precipitation** reaction.

C4 Metal Structure and Properties

Iron and Copper

Iron and copper are two transition metals that have many uses, for example:

- iron is used to make steel, which is used to make cars and girders because it is **very strong**
- copper is used to make electrical wiring as it is a **good conductor**. It is also used to make brass.

Look at the data about metals in this table:

Metal	Relative Electrical Conductivity	Density (g/cm^3)	Relative Hardness
Zinc	1.7	7.1	2.5
Copper	6.1	9.0	3.0
Nickel	1.4	8.9	4.0
Cobalt	1.7	8.9	5.0

You could be asked to pick out the hardest metal (i.e. cobalt) or explain why copper is used in electrical wires (i.e. it is the best electrical conductor).

Physical Properties

Metals are very useful materials because of their properties. They:

- are **lustrous** (shiny), e.g. gold is used in jewellery
- are **hard** and have a **high density**, e.g. steel is used to make drill bits
- have high **tensile strength** (able to bear loads), e.g. steel is used to make girders
- have **high melting** and **boiling points**, e.g. tungsten is used to make light bulb filaments
- are **good conductors** of heat and electricity, e.g. copper is used to make saucepans and wiring.

Metal Structure

Metal atoms are held together by metallic bonds and are packed very closely in a regular arrangement.

Metals have high melting and boiling points because lots of energy is needed to overcome the strong metallic bonds. As the metal atoms pack together, they build a structure of **crystals**.

HT Metal atoms are packed so close together that the outer electron shells **overlap** and form **metallic bonds**. The overlap allows electrons to move about freely. The structure can be described as closely packed metal ions in a 'sea' of **delocalised** (free) electrons.

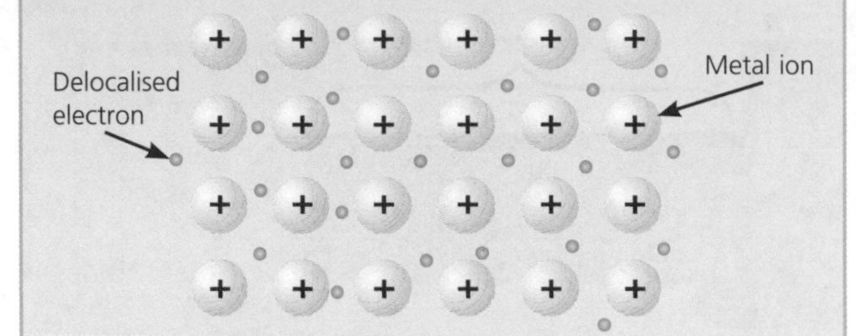

The free movement of the delocalised electrons allows the metal to conduct electricity (e.g. in wiring). The metal is held together by strong forces (the **electrostatic attraction** between the metal ions and the delocalised electrons). This is why many metals have high melting and boiling points.

Superconductors

Metals are able to conduct electricity because the atoms are very close together and the electrons can move from atom to atom.

At low temperatures, some metals can become **superconductors**. A superconductor has very little, or no, resistance to the flow of electricity. Very low resistance is useful when you require:

- a powerful electromagnet, e.g. inside medical scanners
- very fast electronic circuits, e.g. in a supercomputer
- power transmission that does not lose energy.

HT The search is on to find a superconductor that will work at room temperature (20°C). The majority of superconductors currently in use operate at temperatures below -200°C. This very low temperature is costly to maintain and impractical for large-scale uses.

Water

The four main **sources** of water are:

- rivers
- lakes
- reservoirs
- aquifiers (wells and bore holes).

Example

The pie chart shows the sources of water in Northern Ireland.

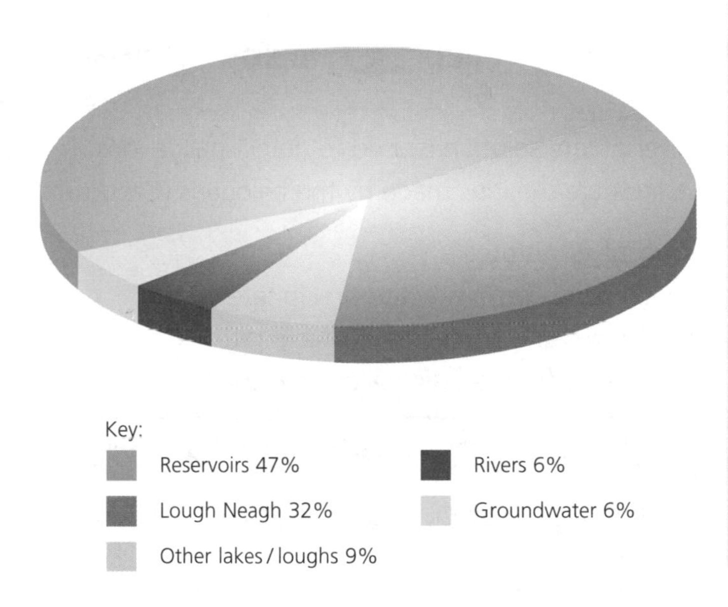

You will not be expected to remember this data, but you may be asked to interpret data like this. For example, using this data you could be asked to pick out the largest water source in Northern Ireland.

Water is an important resource for industry as well as being essential for drinking, washing and other jobs in the home. The chemical industry uses large amounts of water for **cooling**, as a **solvent** and as a cheap **raw material**.

In some parts of Britain the demand for water is higher than the supply, so it is important not to waste water.

Water Treatment

Water has to be **treated** to **purify** it and make sure it is safe to drink before it reaches homes or factories. Untreated (raw) water can contain:

- insoluble particles
- pollutants
- microbes
- dissolved salts and minerals.

A typical treatment process is shown below:

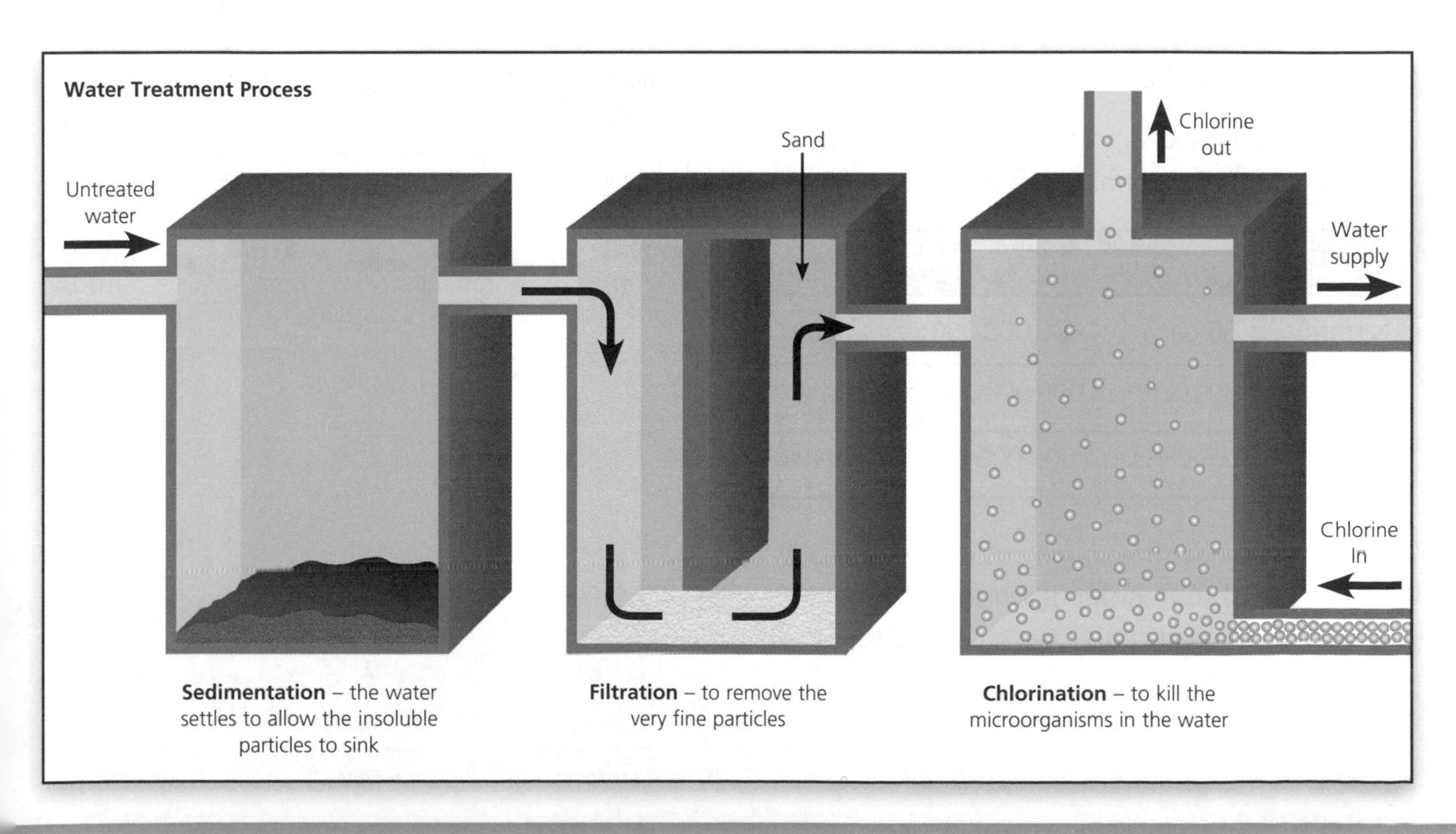

Water Purification

Tap water is not pure. It contains soluble materials that are not removed in the normal water treatment process. There is the possibility that some of the materials may be poisonous and, in that case, extra steps must be taken to remove them. Water must be distilled to make sure it is absolutely pure. This uses lots of energy, so it is very costly.

Huge amounts of expensive energy would be needed in order to distil sea water. And because sea water is quite corrosive, the equipment needed would be very costly. These factors make the cost of making drinking water out of sea water prohibitive, i.e. we are prevented from doing it because the cost is too high.

Pollution in Water

The **pollutants** that may be found in water supplies include nitrates from the run-off of fertilisers, lead compounds from old pipes in the plumbing, and pesticides from spraying crops near to the water supply. These materials are more difficult to remove from the water.

The table below shows some pollutants and the maximum amounts permitted in drinking water.

Pollutant	Maximum Amount Permitted
Nitrates	50 parts in 1 000 000 000 parts water
Lead	50 parts in 1 000 000 000 parts water
Pesticides	0.5 parts in 1 000 000 000 parts water

Again, you do not have to remember the data but you may be asked to pick out which pollutant has the smallest allowed concentration, or transfer the data onto a graph.

Dissolved Ions

The **dissolved ions** of some salts are easy to identify as they will undergo **precipitation** reactions. A precipitation reaction occurs when an insoluble solid is made from mixing two solutions together.

Sulfates can be detected using barium chloride solution: a white precipitate of barium sulfate forms, as in the following example:

sodium sulfate + barium chloride ⟶ barium sulfate (white) + sodium chloride

$$Na_2SO_4(aq) + BaCl_2(aq) \longrightarrow BaSO_4(s) + 2NaCl(aq)$$

Silver nitrate solution is used to detect halide ions. Halides are the ions made by the halogens (Group 7).

With silver nitrate:
- chlorides form a white precipitate
- bromides form a cream precipitate
- iodides form a pale yellow precipitate.

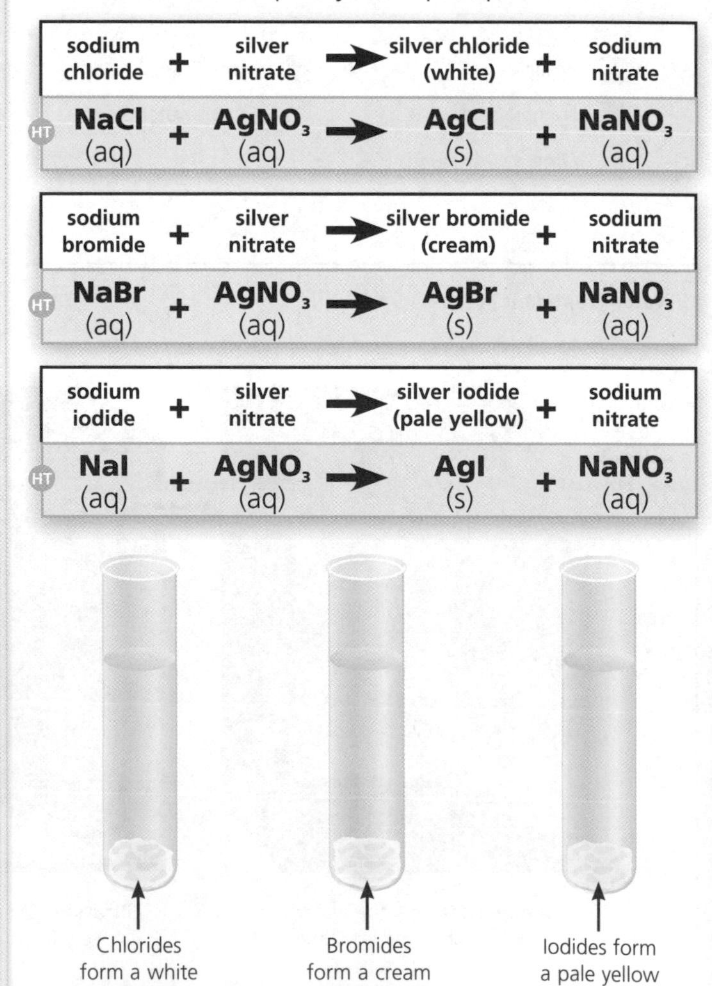

1. **a)** What are isotopes? **[2]**

   b) What is the electronic structure of silicon? (Atomic number = 14) **[1]**

2. **a) i)** Explain how you make a 1^+ ion (e.g. Na^+) from a neutral atom. **[1]**

   ii) Explain how you make a 2^- ion (e.g. O^{2-}) from a neutral atom. **[1]**

   b) Draw the electronic structure of a sodium atom and a sodium ion (atomic number = 11). **[2]**

3. **a)** Name the gas that is used to make plastics and pesticides and is also used to sterilise water. **[1]**

   b) Write the word equation to show the reaction of sodium iodide solution with chlorine gas. **[1]**

4. **a)** Explain why iron and steel are so useful for making cars and bridges. **[1]**

   b) Gold and silver jewellery are lustrous. What does this mean? **[1]**

   c) Suggest one other property that is desirable in a metal used to make jewellery. **[1]**

5. Explain why a silicon atom has no overall charge. **[1]**

6. Explain why ionic compounds such as sodium chloride have such high melting points and will conduct electricity when molten, but covalent compounds, such as methane, have low melting points and do not conduct electricity. **[6]**

   The quality of your written communication will be assessed in your answer to this question.

7. **a) i)** Write the formula equation to show the reaction of sodium iodide solution with chlorine gas. **[2]**

   ii) What is the name of this type of reaction? **[1]**

   iii) Use the reaction to compare the reactivity of chlorine and iodine. **[1]**

   b) Write the ionic equation to show the formation of bromide ions from a bromine molecule and explain why the reaction is reduction. **[3]**

8. **a)** What is a superconductor? **[1]**

   b) Describe one benefit and one problem of using a superconductor. **[2]**

9. **a)** Suggest which of the following is not a transition metal. Use a Periodic Table to help you. **[1]**

   Copper **Iron** **Zinc** **Silver** **Calcium** **Gold**

HT

9. **b)** Complete this equation
   $Fe^{2+} + OH^- \longrightarrow$ **[2]**

10. **a)** Draw the electronic structure of a magnesium ion and a chloride ion (atomic numbers: Mg = 12; Cl = 17). **[2]**

    b) Use your answer to part (a) to work out the formula of magnesium chloride. **[1]**

11. Draw a dot and cross diagram to show the covalent bonding in methane, CH_4. **[2]**

Moles and Molar Mass

C5: How Much? (Quantitative Analysis)

This module looks at:

- Molar mass and the mole, and related calculations.
- Empirical formulae and calculating percentage composition.
- Volumes, calculating concentrations and GDAs on food labels.
- Carrying out titrations and processing the results, indicators and pH changes.
- Measuring gas volumes produced in reactions and processing the results.
- Reversible reactions, equilibria and the effect of changing conditions.
- Strong and weak acids: properties, reactions and differences.
- Precipitation reactions to identify halides and sulfates, and ionic equations.

Molar Mass

The amount of substance in a chemical reaction is measured in **moles**.

The mass of one mole (the **molar mass**) of any substance is the **relative formula mass** (***M_r***) in grams (g). The unit for molar mass is g/mol.

To work out the mass of one mole of a substance, calculate the relative formula mass (see page 44) and then add the unit 'g' to the answer.

Example 1

What is the molar mass of sodium hydroxide (NaOH)?

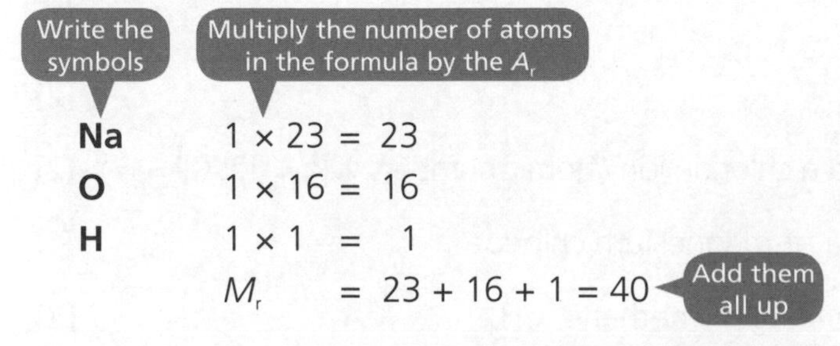

The M_r of NaOH is 40 and so the mass of 1 mole of NaOH is **40g**. So, the molar mass is **40g/mol**.

Example 2

What is the molar mass of magnesium hydroxide ($Mg(OH)_2$)?

Mg	1 × 24	= 24
O	2 × 16	= 32
H	2 × 1	= 2
	M_r	= 24 + 32 + 2 = 58

The M_r of $Mg(OH)_2$ is 58 and so the mass of 1 mole of $Mg(OH)_2$ is **58g**. So, the molar mass is **58g/mol**.

Conservation of Mass

During a chemical reaction, no mass is lost or gained, i.e. it is **conserved**. For example, when 100g of calcium carbonate decomposes, it produces 56g of calcium oxide and 44g of carbon dioxide:

calcium carbonate	⟶	calcium oxide	+	carbon dioxide
$CaCO_3(s)$	⟶	$CaO(s)$	+	$CO_2(g)$

100g ⟶ 56g + 44g = 100g

100g reactants = 100g products

However, the mass measured at the end of a reaction might be:

- greater if a gas has been gained from the air **or**
- smaller if water vapour or a gas has been allowed to escape.

Example

When 50g of calcium carbonate is heated in a thermal **decomposition** reaction, 28g of calcium oxide is made. What mass of carbon dioxide is lost?

calcium carbonate	⟶	calcium oxide	+	carbon dioxide
$CaCO_3(s)$	⟶	$CaO(s)$	+	$CO_2(g)$

50g ⟶ 28g + mass of carbon dioxide

Mass of carbon dioxide = 50g – 28g = **22g**

Reacting Ratios

If the reacting masses in a reaction are known, further reacting masses can be calculated using simple **ratios**.

Example

The reaction between 159.5g of copper sulfate and 106g of sodium carbonate produces 123.5g of copper carbonate and 142g of sodium sulfate:

$$CuSO_4 + Na_2CO_3 \longrightarrow CuCO_3 + Na_2SO_4$$

159.5g + 106g ⟶ 123.5g + 142g

a) How much copper sulfate and sodium carbonate are needed to produce 370.5g of copper carbonate?

370.5 is 3 × 123.5g, so you just need to multiply all the above masses by the same amount (×3) to find the new set of masses.

3 × 159.5g + 3 × 106g ⟶ 3 × 123.5g + 3 × 142g
= 478.5g + 318g ⟶ 370.5g + 426g

So, **478.5g** of copper sulfate and **318g** of sodium carbonate are needed.

b) How much copper sulfate and sodium carbonate are needed to produce 12.35g of copper carbonate?

12.35g is 123.5g ÷ 10, so you just need to divide all the above masses by the same amount (÷ 10) to find the new set of masses.

$$\frac{159.5g}{10} + \frac{106g}{10} \longrightarrow \frac{123.5g}{10} + \frac{142g}{10}$$

= 15.95g + 10.6g ⟶ 12.35g + 14.2g

So, **15.95g** of copper sulfate and **10.6g** of sodium carbonate are needed.

N.B. A quick way to check that your calculation is correct is to add the mass of the reactants together and see if the total is equal to the total mass of the products.

HT Molar Mass, Moles and Mass

The relative atomic mass of an element is the average mass of the atoms of that element compared with a twelfth of the mass of a carbon-12 (^{12}C) atom.

The number of moles of an element or a compound can be calculated using this formula:

$$\text{Number of moles} = \frac{\text{Mass}}{\text{Molar mass}}$$

Example 1

How many moles of iron are there in 112g? The A_r of iron is 56.

Number of moles $= \frac{112g}{56g} =$ **2 moles**

Example 2

How many moles of ethanol are there in 230g? The M_r of ethanol is 46.

Number of moles $= \frac{\text{Mass}}{\text{Molar mass}}$

$= \frac{230g}{46g} =$ **5 moles**

Example 3

What is the mass of oxygen in 3 moles of aluminium oxide (Al_2O_3)?

3 moles of aluminium oxide contains 3 × 3 = 9 moles of oxygen.

Mass = Number of moles × Molar mass
= 9 moles × Molar mass of oxygen (O)
= 9 × 16g = **144g**

Moles and Reacting Masses

With a formula equation and the relative atomic masses you can work out the mass of a product or reactant in a reaction.

Example

What mass of aluminium do you get when you electrolyse 408g of aluminium oxide?
(A_r of Al = 27, A_r of O = 16)

$2Al_2O_3 \rightarrow 4Al + 3O_2$

M_r of Al_2O_3 = (2 × 27) + (3 × 16) = 102

Number of moles of $Al_2O_3 = \frac{\text{Mass}}{\text{Molar mass}} = \frac{408}{102}$
= **4 moles**

From the equation, 2 moles of Al_2O_3 make 4 moles of Al. So, 4 moles of Al_2O_3 will make 2 × 4 = 8 moles of Al. 8 moles of Al = 8 × 27 = 216g of aluminium.

C5 Percentage Composition and Empirical Formula

Mass of Elements in a Compound

The mass of a compound is made up of all the masses of its elements added together. Therefore, if the mass of a compound and the mass of one of the elements is known, the mass of the other element can be calculated.

Example
79.5g of copper oxide contains 16g of oxygen. What mass of copper does it contain?

Mass of copper + 16g = 79.5g
Mass of copper = 79.5g – 16g = **63.5g**

Empirical Formula

The **empirical formula** gives the simplest whole number ratio of each type of atom in a compound.

The formulae of the first three alkenes are ethene C_2H_4, propene C_3H_6, butene C_4H_8. All three alkenes have the same empirical formula C_1H_2.

The formula of ethanoic acid is CH_3COOH. You should be able to work out that the empirical formula of ethanoic acid is CH_2O.

Percentage Composition

The percentage mass of an element in a compound can be calculated from experimental data. The information you need is the total mass of the compound and the mass of the element that it contains.

$$\text{Percentage mass} = \frac{\text{Mass of the element}}{\text{Mass of the compound}} \times 100$$

Example
What is the percentage mass of nitrogen in 8.0g of ammonium nitrate that was found to contain 2.8g of nitrogen?

$$\text{Percentage mass} = \frac{\text{Mass of the element}}{\text{Mass of the compound}} \times 100$$

$$= \frac{2.8}{8.0} \times 100 = \mathbf{35\%}$$

The percentage composition of an element in a compound can be calculated from its formula – the only other information you need is the relative atomic masses of the elements involved.

Example
Calculate the percentage composition of iron in iron(III) oxide, Fe_2O_3. (A_r iron = 56; A_r oxygen = 16)
M_r Fe_2O_3 = (2 × 56) + (3 × 16) = 160
Iron represents (2 × 56) = 112 of this.

$$\text{Percentage of iron} = \frac{\text{Mass of the element}}{\text{Mass of the compound}} \times 100$$

$$= \frac{112}{160} \times 100 = \mathbf{70\%}$$

HT Calculating Empirical Formulae

The empirical formula of a compound can be calculated from either:

- the percentage composition of the compound by mass **or**
- the mass of each element in the compound.

The following method is used to calculate empirical formulae:

1. List all the elements in the compound.
2. Divide the data for each element by its relative atomic mass.
3. Select the smallest answer from step 2 and divide each answer by that result to obtain a ratio.
4. The ratio may have to be scaled up or down to give whole numbers.

Example
What is the empirical formula of a hydrocarbon containing 75% carbon? (Hydrogen is the other element = 25%.)

Carbon : Hydrogen

$$\frac{75}{12} : \frac{25}{1}$$

= 6.25 : 25

6.25 : 25
÷ 6.25 ÷ 6.25
= 1 : 4

6.25 is the smallest result so divide both sides by 6.25

So, the empirical formula is C_1H_4, or $\mathbf{CH_4}$

Volume

The two units commonly used to measure the **volume** of liquids and solutions are:

- cm^3 (cubic centimetres)
- dm^3 (cubic decimetres); $1dm^3$ is also known as 1 litre.

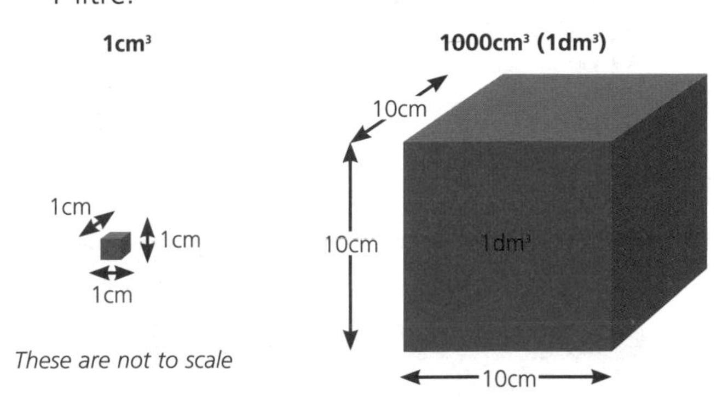

To convert a volume:

- from cm^3 into dm^3, divide it by 1000
- from dm^3 into cm^3, multiply it by 1000.

Example

Convert $2570cm^3$ into dm^3.

$\frac{2570}{1000} = \mathbf{2.57dm^3}$

Concentration

The **concentration** of a solution can be measured in:

- g/dm^3 (grams per cubic decimetre)
- mol/dm^3 (moles per cubic decimetre).

In a concentrated solution, the solute particles are more crowded together than they are in a **dilute** solution.

There are more solute particles in the same volume.

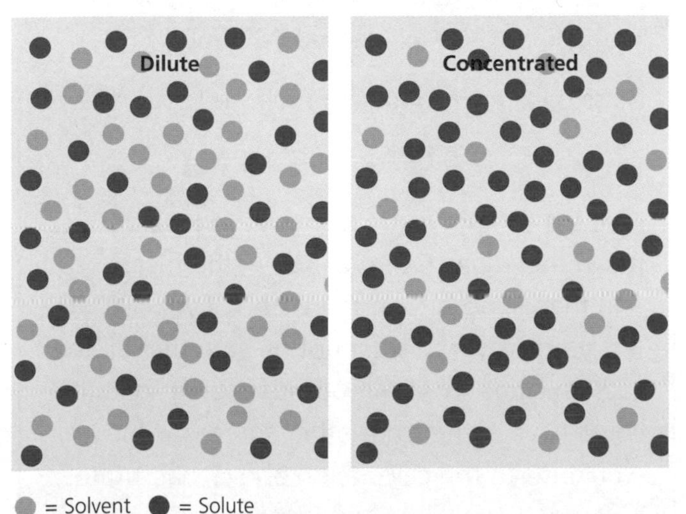

A concentrated solution can be made more dilute by adding water.

The table below shows how much water should be added to a $1mol/dm^3$ solution to produce a $0.1mol/dm^3$ solution:

Volume of $1mol/dm^3$ Solution	Volume of Water Added	Volume of $0.1mol/dm^3$ Solution
$1cm^3$	$9cm^3$	$10cm^3$
$10cm^3$	$90cm^3$	$100cm^3$
$100cm^3$	$900cm^3$	$1000cm^3$

Some concentrated solutions must be diluted before they are used – for example, orange cordial has to be diluted before it is drunk, otherwise it would taste too strong.

It is important to accurately follow the instructions on products such as baby milk and dilute them correctly. If the concentration is wrong then it may harm the baby. If a medicine is too dilute, it may not work properly; if too concentrated it could lead to an overdose and make you more ill.

HT Concentrations in mol/dm³

The concentration of a solution in mol/dm³ can be calculated using the following formula:

$$\text{Concentration (mol/dm}^3\text{)} = \frac{\text{Amount of solute (moles)}}{\text{Volume of solvent (dm}^3\text{)}}$$

Example

What is the concentration of the solution (in mol/dm³) when 5 moles of copper chloride is dissolved in 10dm³ water?

$$\text{Concentration} = \frac{\text{Amount of solute}}{\text{Volume of solvent}}$$

$$\text{Concentration} = \frac{5}{10}$$

$$= \mathbf{0.5mol/dm^3}$$

You should be able to use the equation rearranged:

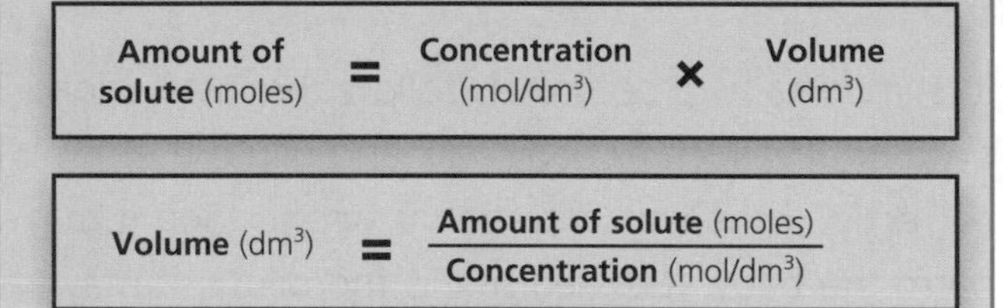

$$\text{Amount of solute (moles)} = \text{Concentration (mol/dm}^3\text{)} \times \text{Volume (dm}^3\text{)}$$

$$\text{Volume (dm}^3\text{)} = \frac{\text{Amount of solute (moles)}}{\text{Concentration (mol/dm}^3\text{)}}$$

Food Labels and GDAs

Ingredients on a food label are always listed in order of mass with the greatest first. The label below shows that the ingredient with the largest mass is wheat flakes and the smallest amount is Vitamin B12.

The **guideline daily amount** (GDA) is a guide to how much of a food ingredient is needed by a person per day for a healthy diet. In your exam, you may be asked to read the guideline daily amount from a food label such as the one shown here.

Vitamins and Minerals	In a 30g serving	%GDA
Salt	0.4g	10%
Vitamin D	1.5μg	30%
Iron	2.4mg	17%
Folic acid	68μg	34%
Niacin	5.5mg	31%
Vitamin B_{12}	0.5mg	34%

Example

A 30g serving of the cereal provides 10% of the GDA of salt. What is the GDA of salt?

0.4g salt = 10% GDA. So, 0.4g × 10 = **4g**

HT Calculating GDAs

Food labels have to be looked at carefully because they can be misleading. For example, the label of a microwave meal states that it contains 2g of sodium. If all the sodium came from salt (sodium chloride), you could calculate the mass of salt (M_r = 58.5) that contains 2g of sodium (A_r = 23).

$$\text{Number of moles} = \frac{\text{Mass}}{\text{Molar mass}}$$

$$= \frac{2}{23} = 0.087\text{mol}$$

$$\text{Mass of salt} = \text{Number of moles} \times \text{Molar mass}$$

$$= 0.087 \times 58.5 = \mathbf{5.09g}$$

If all the sodium came from salt, it would mean there was 5.09g of salt in the meal, and that is more than the GDA (GDA salt = 4g). However, we cannot be sure that all the sodium came from salt because some sodium could have come from other sources, which would make the calculation above inaccurate.

Titration

A **titration** is a method that can be used to find the volume of acid needed to exactly neutralise an alkali (or the other way round). The amount needed depends on the concentration of the acid used. This is how to carry out a titration:

1. Using a pipette and filler, measure the alkali into a conical flask. Using a filler is the safe way to draw alkali into a pipette.
2. Add a few drops of indicator to the flask.
3. Fill the burette with acid.
4. Record the start volume of acid.
5. Add acid slowly to the alkali until the indicator just changes colour (the end point).
6. Record the end volume of acid.
7. Calculate how much acid has been added (final volume – start volume). This is called the **titre**.
8. To check the accuracy of the results, repeat the titration until you have consistent results.

In some circumstances, the alkali can be put in the burette, and the acid in the conical flask.

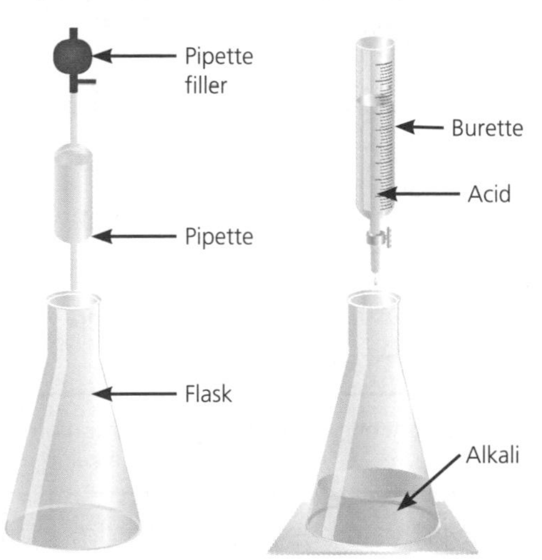

Titration Results

Titration data should be collected in a table such as this:

Experiment	1	2	3
Start volume cm^3	2.5	1.7	2.1
Final volume cm^3	20.8	20.1	20.5
Titre cm^3 (final volume – start volume)	18.3	18.4	

You should be able to calculate the missing titre. ($20.5 - 2.1 = 18.4$ cm^3). The titration is repeated a number of times to get consistent titres so that the average titre used for calculation is as accurate as possible.

Indicators

The table below shows the colours that **indicators** turn when they are in acidic and alkaline solutions:

Indicator	Colour in Acid	Colour in Alkali
Litmus	Red	Blue
Phenolphthalein	Colourless	Pink
Universal indicator	Red	Blue

Single indicators, such as litmus or phenolphthalein, produce a sudden, sharp colour change during titration, which clearly shows the end point.

Universal indicator is a mixture of different indicators, which gives a continuous range of colours. The pH of a solution can be determined by comparing the colour of the indicator in solution to a pH colour chart. This is less suitable for titrations.

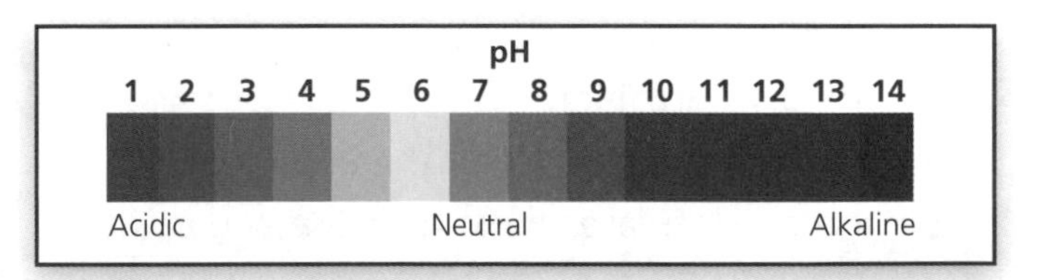

HT In an acid–base titration, it is better to use a single indicator, such as litmus, to determine the end point because the pH changes very suddenly near the end point. The end point is much more confused when a mixed indicator, such as universal indicator, is used because it gives a range of colours.

pH Curves

pH curves can be drawn to show what happens to the pH when an acid is added to an alkali, or vice versa:

- An acid has a low pH. When an alkali is added to it, the pH increases (see Graph 1 on page 76).
- An alkali has a high pH. When an acid is added to it, the pH decreases (see Graph 2 on page 76).

pH Curves (cont)

Whether an acid is being added to an alkali, or an alkali is being added to an acid, the pH changes very suddenly at the end point of the reaction.

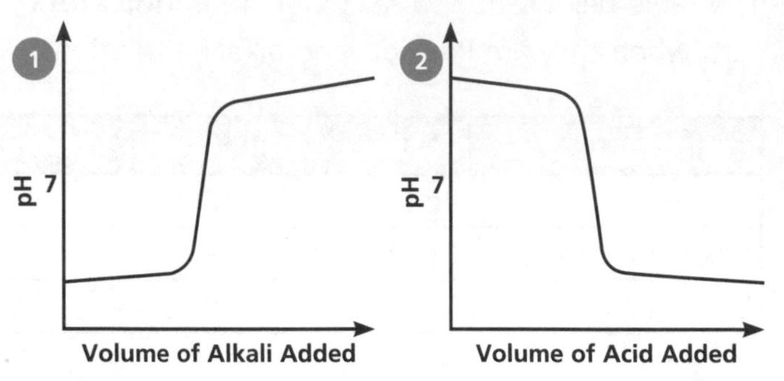

Example

The graph opposite shows how the pH of a solution changes as an acid is added to it.

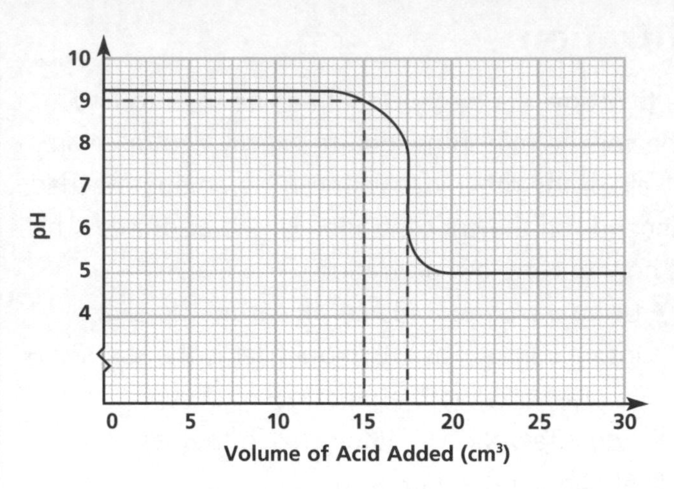

a) What was the pH when $15cm^3$ acid had been added?
The pH was 9.

b) How much acid was added to reach the end point?
$17.5cm^3$ acid was added.

HT When sketching a pH curve starting with an acid:

- the pH starts low
- the pH increases when the alkali is added
- the pH increases most near the end point.

When sketching a pH curve starting with an alkali:

- the pH starts high
- the pH decreases when the acid is added
- the pH decreases most near the end point.

Using Concentration Formulae

The following formulae are used to calculate concentration, volume and moles:

$$\textbf{Concentration}\ (mol/dm^3) = \frac{\textbf{Moles}}{\textbf{Volume}\ (dm^3)}$$

$$\textbf{Volume}\ (dm^3) = \frac{\textbf{Moles}}{\textbf{Concentration}\ (mol/dm^3)}$$

$$\textbf{Moles} = \textbf{Concentration}\ (mol/dm^3) \times \textbf{Volume}\ (dm^3)$$

Example 1

What is the volume of a $4mol/dm^3$ solution that will add 0.2 moles to a mixture?

$$\text{Volume} = \frac{\text{Moles}}{\text{Concentration}} = \frac{0.2}{4} = \mathbf{0.05dm^3}$$

Example 2

What is the number of moles added by $0.25dm^3$ of a $2mol/dm^3$ solution?

Moles = Concentration × Volume
= 2 × 0.25 = **0.5 moles**

At the end point of a titration in which the acid and alkali react in a one-to-one ratio, the number of moles of acid is equal to the number of moles of alkali:

$$\textbf{Concentration of acid} \times \textbf{Volume of acid} = \textbf{Concentration of alkali} \times \textbf{Volume of alkali}$$

Convert volumes in cm^3 into dm^3 before you use this equation.

Example

$0.025dm^3$ of a sample of an alkali is completely neutralised by $0.03dm^3$ of a $0.1mol/dm^3$ acid. What is the concentration of alkali?

$$\text{Concentration of acid} \times \text{Volume of acid} = \text{Concentration of alkali} \times \text{Volume of alkali}$$

0.1 × 0.03 = Concentration of alkali × 0.025

$$\text{Concentration of alkali} = \frac{0.1 \times 0.03}{0.025}$$

$$= \mathbf{0.12mol/dm^3}$$

Measuring Gas Volumes

The following apparatus can be used to collect and measure the **volume** of a gas made in a reaction:

- An upturned measuring cylinder.

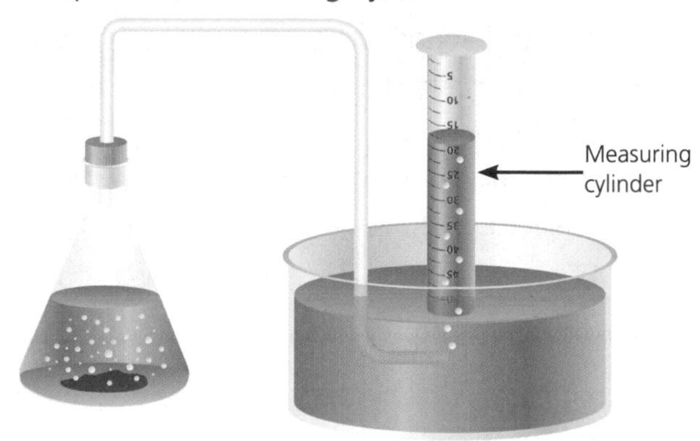

- A gas syringe.

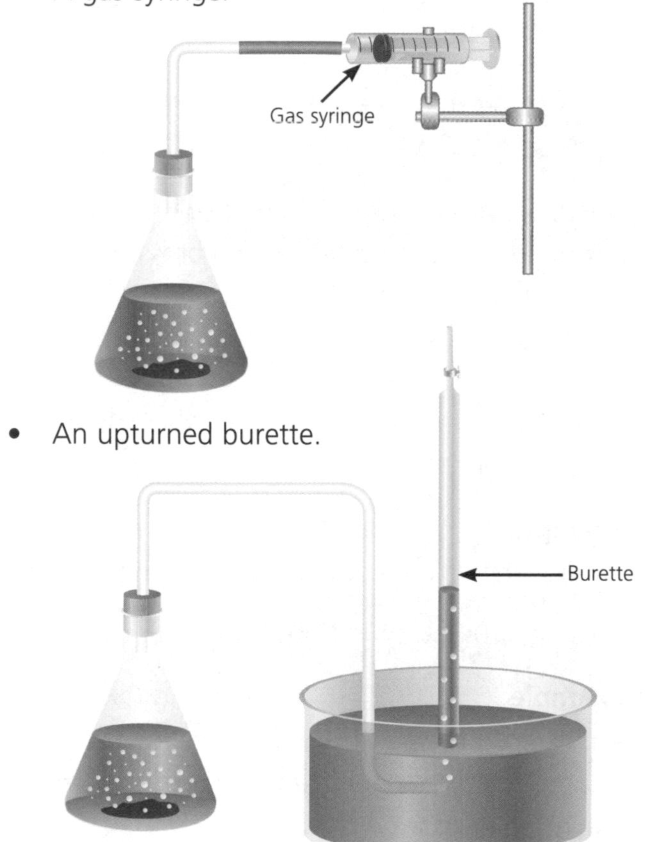

- An upturned burette.

The following method should be used to measure the volume of gas produced in an experiment:

1. Measure out the reactants.
2. Add the reactants together in the flask and start the stopclock or timer.
3. Measure and record the volume of gas produced at regular time intervals until the volume stops increasing. (This indicates the reaction has finished.)

Measuring Gas Masses

The amount of gas produced in a reaction can also be measured by monitoring the change in mass of a reaction.

The **mass** of a gas made in a reaction can be measured using the following apparatus:

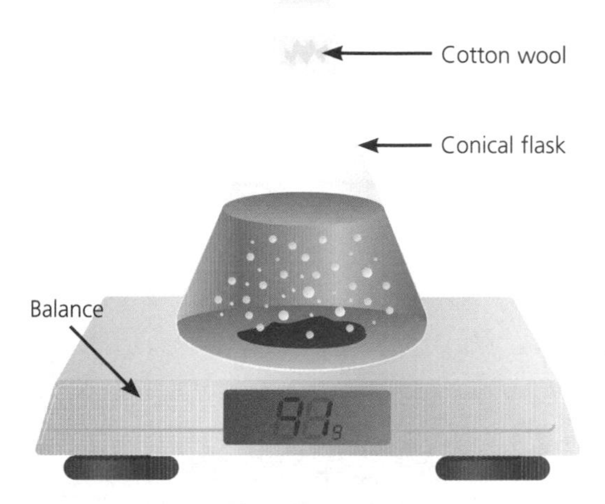

The following method should be used to measure the mass of gas produced in an experiment:

1. Weigh an empty conical flask.
2. Measure out the reactants.
3. Calculate the total mass of the reactants and the flask.
4. Add the reactants together in the flask and start the stopclock or timer.
5. Read and record the mass of the flask and reactants at regular time intervals, until the mass stops decreasing. (This indicates the reaction has finished.)

End of a Reaction

The more reactant that is used, the greater the amount of product (in this case, gas) produced. A reaction stops when one of the reactants has been used up. The reactant that gets used up first is called the **limiting reactant**.

In a reaction in which there is a one-to-one ratio between reactants, the limiting reactant is the one with the smallest number of moles.

The amount of gas produced in a reaction is directly proportional to the amount of limiting reactant used.

Gas Volumes

Sketch Graphs of Reactions

The results recorded for a reaction can be used to produce a graph, for example:

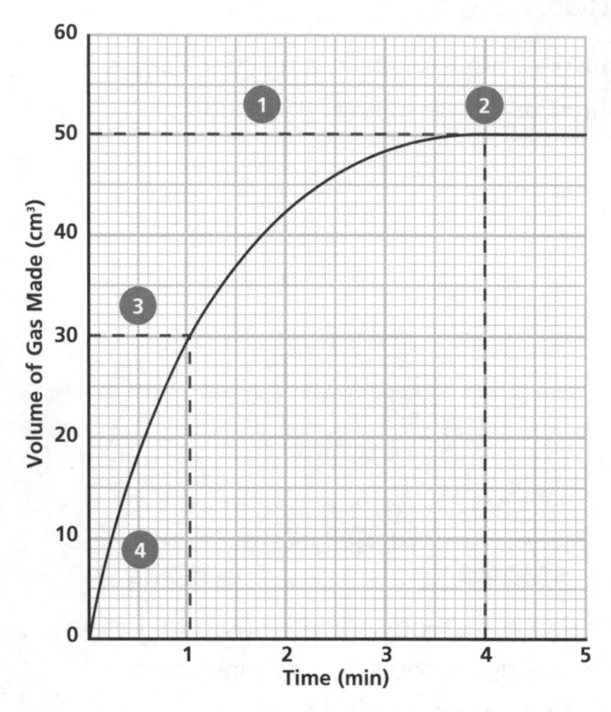

The following information can be read from the graph:

1. The total volume of gas produced.
2. When the reaction ended.
3. The volume of gas produced at a particular time (or the time at which a particular volume of gas was produced).
4. The point at which the reaction was fastest.

If a reaction is repeated with differing amounts of limiting reactant, the graphs would look like this:

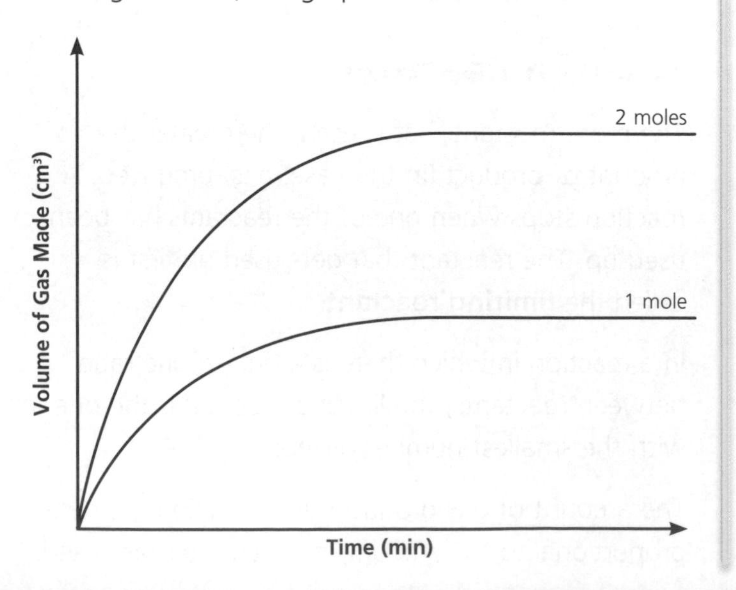

HT When sketching a graph to show the volume of gas made in a reaction, remember the following information:

- The curve should be steepest at the beginning of the reaction (when the rate is fastest).
- The curve should get less steep as the reaction progresses.
- The curve should become horizontal to show the end of the reaction, which should be level with the final volume of gas produced.

The amount of the limiting reagent you start with dictates the amount of gas you will make. When there are more reactant particles, there is more reaction and more gas particles are then produced.

Calculating Volumes and Amounts of Gases

One mole of any gas occupies a volume of **24dm³** at room temperature and pressure. This fact can be used to:

- calculate the volume of a known amount of gas
- calculate the amount of gas if the volume is known.

Example 1

What is the volume of half a mole of nitrogen?

Volume $= 0.5 \times 24 =$ **12dm³**

Example 2

A balloon is filled with oxygen until it has a volume of 6dm³. How many moles of oxygen are in the balloon?

Amount $= \frac{6}{24} =$ **0.25mol**

Reversible Reactions

A **reversible reaction** can go forwards or backwards, in many cases, under the same conditions (i.e. at the same time). It is represented by $\rightleftharpoons$. For example, the reaction between nitrogen and hydrogen to produce ammonia is reversible:

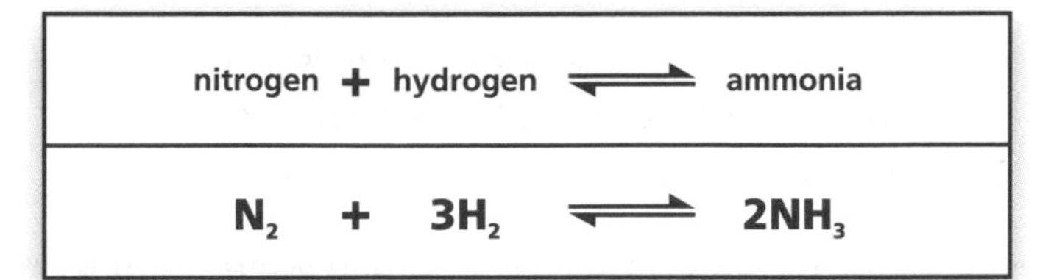

Equilibrium

A reversible reaction can reach **equilibrium** (a balance), i.e. the rate of the forward reaction equals the rate of the backward reaction. At equilibrium, the amounts and concentrations of reactants and products stay the same, even though reactions are still taking place.

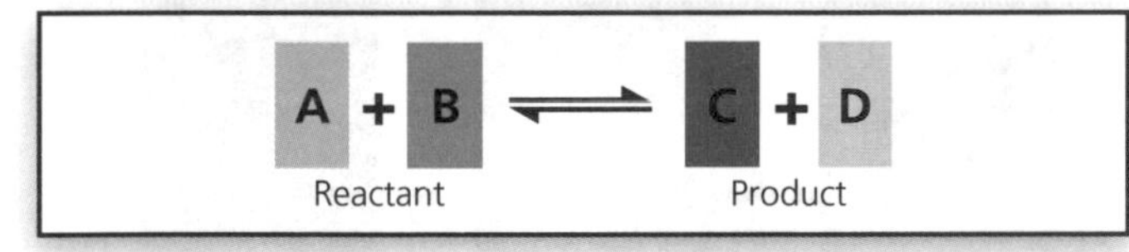

The position of an equilibrium can be altered by changing:

- the temperature
- the pressure
- the concentration of reactant(s) and / or product(s).

If the position of the equilibrium lies to the right of the reaction equation, the concentration of the products is greater than the concentration of the reactants:

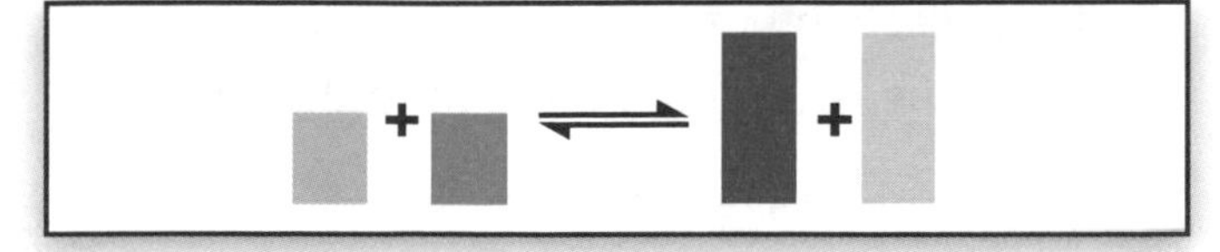

If the position of the equilibrium lies to the left of the reaction equation, the concentration of the products is less than the concentration of the reactants:

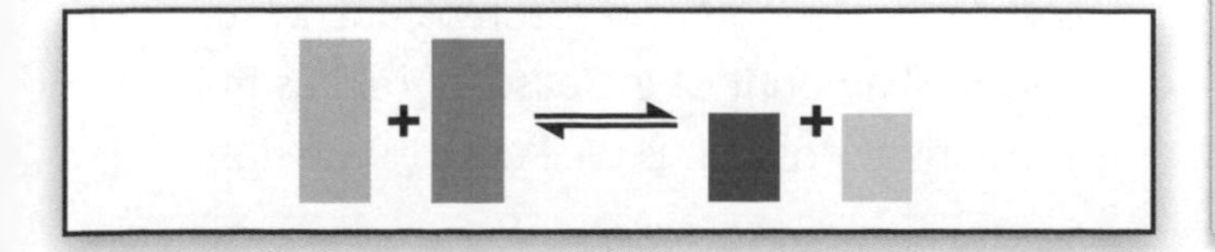

Equilibrium composition can be shown in tables or graphs, and the following information can be obtained:

- the composition at a particular temperature
- the composition at a particular pressure
- the effect of temperature and pressure on composition.

Example

The table and graphs below show how altering the reaction conditions can change the equilibrium composition for the reaction to make ammonia:

Pressure (atmospheres)	Ammonia made at 300°C (%)	Ammonia made at 600°C (%)
100	43	4
200	62	12
300	74	18
400	79	19
500	80	20

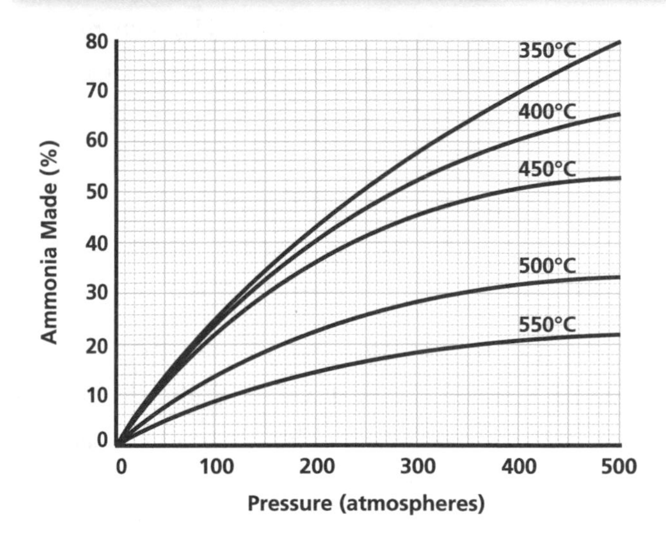

a) What happens to the percentage of ammonia made when the temperature increases?
It falls.

b) What is the percentage of ammonia made at 200 atmospheres and 600°C?
12%.

Use the table

c) What percentage of ammonia is made at 300 atmospheres and 400°C?
52%.

Use the graph

Reversible Reactions and Equilibrium

A reversible reaction will only reach equilibrium if the conditions (such as temperature and pressure) are not changed and no substance is added or removed. This is known as a **closed system**.

As the reaction gets nearer to equilibrium, the forward reaction will slow down and the backward reaction will speed up until both reactions are at the same rate.

The equilibrium can be moved to the right of the reaction equation by:

- adding more reactants **or**
- removing the product as it is made.

The equilibrium can be moved to the left of the reaction equation by:

- reducing the amount of reactants **or**
- increasing the amount of product.

The equilibrium position is affected by a change in temperature. The reaction will be exothermic in one direction and equally endothermic in the other direction. An increase in temperature pushes the equilibrium position in the endothermic direction. A decrease in temperature moves the reaction in the exothermic direction.

In a reaction that involves gases, an increase in pressure moves the equilibrium in the direction that has the smallest number of moles of gas.

The Contact Process

The raw materials sulfur, air and water are made into sulfuric acid in the **Contact Process**:

1. Sulfur is burned in a furnace to make sulfur dioxide:

sulfur + oxygen ⟶ sulfur dioxide

$$S + O_2 \longrightarrow SO_2$$

2. The sulfur dioxide combines with oxygen from the air in a reversible reaction to make sulfur trioxide:

sulfur dioxide + oxygen ⇌ sulfur trioxide

$$2SO_2(g) + O_2(g) \rightleftharpoons 2SO_3(g)$$

The reaction takes place using a vanadium(V) oxide (V_2O_5) **catalyst**, at a temperature of 450°C and at atmospheric pressure.

3. Sulfur trioxide is then dissolved in water to make sulfuric acid:

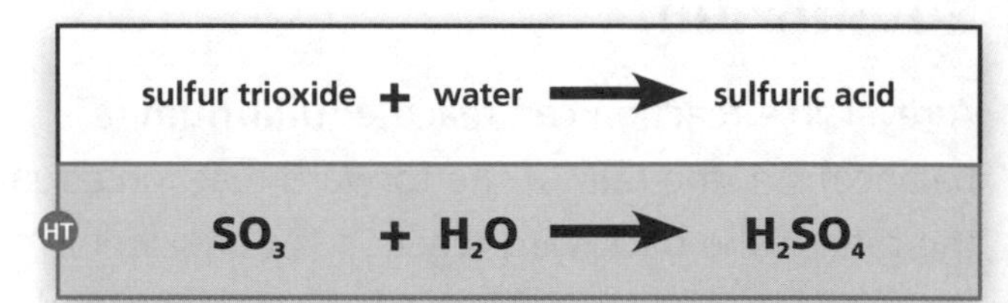

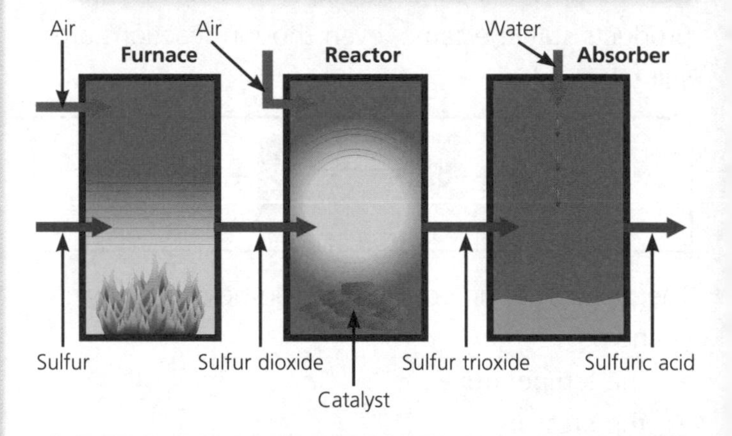

A catalyst is used in the reaction to speed up the rate of production of sulfur trioxide but it does not change the position of the equilibrium.

Increasing the temperature increases the rate of the reaction but it also reduces the yield and pushes the equilibrium position to the left.

So a **compromise** temperature, 450°C, is used to get a balance between yield and rate.

A higher pressure would push the equilibrium to the right and increase the yield, but the extra cost of increasing the pressure is not worth the small amount of increase in yield as the equilibrium position is well over to the right.

Strong and Weak Acids

An acid is a substance that **ionises** in water to produce hydrogen ions (H^+):

- A **strong** acid ionises **completely** in water.
- A **weak** acid only **partially** ionises in water. The ionisation of a weak acid is a reversible reaction and so an equilibrium mixture is made.

Strong acids, such as hydrochloric acid, nitric acid and sulfuric acid, have a lower pH than weak acids, such as ethanoic acid, if they are of the same concentration. Weak acids can be used to flavour food (e.g. vinegar) and to descale kettles and boilers. A strong acid cannot be used to descale a kettle because it may react with the kettle itself.

Reactions with Acids

Ethanoic acid or hydrochloric acid react with:

- magnesium to produce hydrogen gas
- calcium carbonate to produce carbon dioxide gas.

If an equal amount of ethanoic acid and hydrochloric acid are used in the above reactions, the same volumes of gas will be made.

The volume of gas made is determined by the *amount* of reactants used, not by the acid's strength. However, the reaction with ethanoic acid is slower as there are fewer hydrogen ions in it than in the same concentration of hydrochloric acid, and so there are fewer collisions.

Electrolysis of Acids

Acids conduct electricity. A strong acid, such as hydrochloric acid, is a better conductor than the same concentration of a weak acid, such as ethanoic acid. This is because there are fewer hydrogen ions in ethanoic acid to carry the charge.

When hydrochloric acid or ethanoic acid are used in electrolysis, hydrogen gas is made at the negative **electrode**. This is because when electricity is passed through the acid, the hydrogen ions are attracted to the negative electrode, where they are changed into hydrogen gas.

HT More on Strong and Weak Acids

A strong acid produces more H^+ ions than a weak acid of the same concentration because the weak acid does not ionise completely:

- Hydrochloric acid ionises completely:

$$HCl \longrightarrow H^+ + Cl^-$$

- Ethanoic acid ionises partially:

$$CH_3COOH \rightleftharpoons CH_3COO^- + H^+$$

An acid with more H^+ ions (i.e. a strong acid) has a lower pH. A weak acid will have a lower concentration of H^+ ions than a diluted strong acid. It will not be as reactive, so it may be more useful because it will be less destructive.

More on Reactions with Acids

The **strength** of an acid is determined by how much it ionises. The **concentration** of an acid is determined by how many moles of the acid are dissolved in $1dm^3$.

Hydrochloric acid reacts quicker than ethanoic acid because:

- hydrochloric acid is a stronger acid than ethanoic acid
- hydrochloric acid contains a higher concentration of hydrogen ions than ethanoic acid does
- the higher concentration of hydrogen ions in hydrochloric acid leads to a higher collision frequency of hydrogen ions.

More on Electrolysis of Acids

The higher the concentration of hydrogen ions in an acid, the greater the electrical conductivity because the ions carry the charge. This is why strong acids are better conductors than weak acids.

Precipitation Reactions

In a solid ionic substance the ions are in fixed positions, but when they dissolve in water they are free to move about.

A **precipitation** reaction occurs when an **insoluble** solid is made by mixing two ionic solutions together. The **precipitate** (the product) is made when ions from one solution collide and react with ions from the other solution.

The following method can be used to make an insoluble compound:

1. Mix the reactant solutions.
2. Filter off the precipitate.

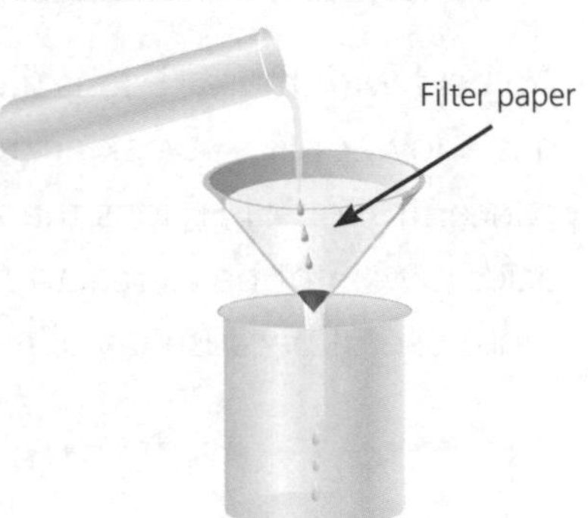

3. Wash the residue in the filter funnel with a little distilled water.
4. Dry the residue (the product) in an oven at 50°C.

Detecting Ions

Halides are the ions made by the halogens (Group 7). Lead nitrate solution can be used to detect halide ions. For example, with lead nitrate:

- chlorides (Cl^-) form a white precipitate
- bromides (Br^-) form a cream precipitate
- iodides (I^-) form a bright yellow precipitate.

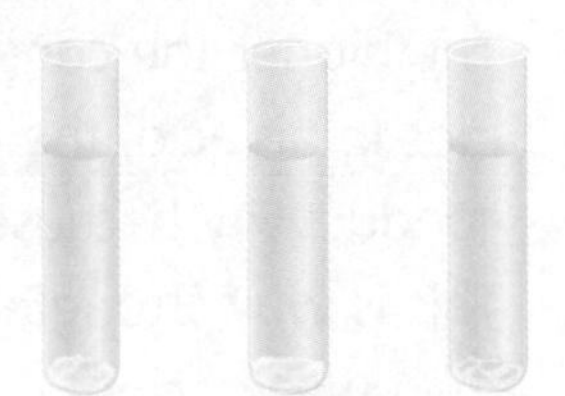

Example word equation:

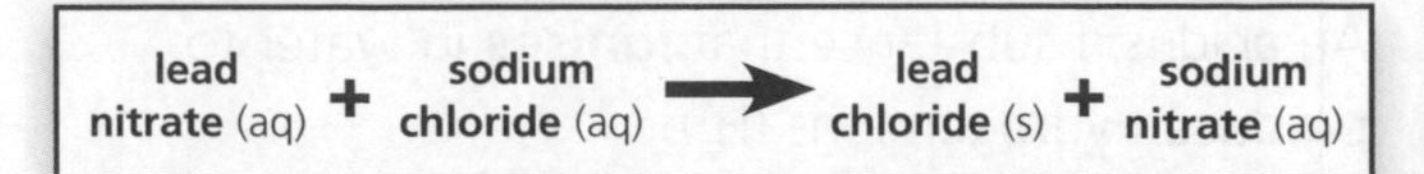

Similar reactions happen when silver ions, Ag^+, are added to a solution of a halide. State symbols are important in a precipitation reaction equation because they show which is the insoluble solid precipitate: (s) – solid. The other state symbols are (l) – liquid, (g) – gas and (aq) – aqueous (i.e. a solution in water).

Sulfate ions can be detected using barium nitrate solution; a white precipitate of barium sulfate forms:

HT Precipitation reactions are very fast because there is a very high frequency of collisions between the ions. For example, when sodium chloride and lead nitrate react, the precipitate of lead chloride forms almost instantly:

$$Pb(NO_3)_2(aq) + 2NaCl(aq) \rightarrow PbCl_2(s) + 2NaNO_3(aq)$$

The ions involved in this reaction are Na^+, Cl^-, Pb^{2+} and NO_3^-.

The Na^+ and the NO_3^- ions stay dissolved in the water and do not do anything. They are called **spectator ions**.

An ionic equation is written by picking out the ions that react to form the precipitate, for example:

$$Pb^{2+}(aq) + 2Cl^-(aq) \rightarrow PbCl_2(s)$$

$$Pb^{2+}(aq) + 2Br^-(aq) \rightarrow PbBr_2(s)$$

$$Pb^{2+}(aq) + 2I^-(aq) \rightarrow PbI_2(s)$$

$$Ba^{2+}(aq) + SO_4^{2-}(aq) \rightarrow BaSO_4(s)$$

Exam Practice Questions

1 a) Look at this reaction: **copper carbonate → copper oxide + carbon dioxide**

Calculate the mass of copper carbonate that must have been thermally decomposed to make 7.95g of copper oxide and 4.4g of carbon dioxide. **[2]**

b) If 63.6g of copper oxide had been made by the above reaction, calculate the mass of carbon dioxide that would have been produced. **[2]**

c) 117g of sodium chloride contains 46g of sodium. What mass of chlorine does it contain? **[2]**

2 a) i) Convert the volume 150cm^3 into dm^3. **[2]**

ii) Convert the volume 2.75dm^3 into cm^3. **[2]**

b) Calculate the concentration of the solution made when 10cm^3 of a 2 mol/dm^3 solution is diluted with water to a total volume of 100cm^3. **[2]**

3 a) Name the type of reaction where an acid reacts with an alkali. **[1]**

b) What is a titre and how is it calculated? **[2]**

4 a) Draw the symbol you would use in an equation to show a reversible reaction. **[1]**

b) Compare the rate of the forward reaction with the rate of the reverse reaction when the overall reaction is at equilibrium. **[1]**

5 Name the three raw materials and the catalyst needed for the Contact process. **[4]**

6 a) Suggest which acid has a higher pH. Choose from 0.1 mol/dm^3 sulfuric acid (strong) or 0.1 mol/dm^3 citric acid (weak). **[1]**

b) Georgie says that 2 mol/dm^3 nitric acid contains more hydrogen ions than 2 mol/dm^3 ethanoic acid. Emily thinks the ethanoic acid has more hydrogen ions. Who is correct? **[1]**

7 a) Describe how you would make a clean, dry sample of lead chloride from a solution of sodium chloride and a solution of lead nitrate. Explain how you would obtain a sample of the other product, sodium nitrate. **[6]**

The quality of your written communication will be assessed in your answer to this question.

b) Write the word equation for the reaction of lead nitrate with sodium chloride. **[1]**

HT

c) Explain why sodium ions can be described as 'spectator ions' in the above reaction. **[1]**

8 The formula of sodium hydroxide is NaOH.

a) Calculate how many moles there are in 16g of sodium hydroxide. **[2]**

b) Calculate the mass of oxygen (A_r=16) in 3 moles of sodium hydroxide. **[2]**

9 A lower temperature gives a higher yield in the Contact Process. Explain why a relatively high temperature of 450°C is used. **[2]**

10 Amerjit reacts magnesium with hydrochloric acid and she makes 4.8dm^3 of hydrogen gas. Calculate the number of moles of gas she has made. **[2]**

C6: Chemistry Out There

This module looks at:

- Reactions using electricity to decompose ionic liquids and solutions.
- Making electricity from a chemical reaction, and fuel cells.
- Redox and displacement reactions, rusting of iron and steel and protection from corrosion.
- Alcohols, making ethanol by fermentation and hydration of ethene.
- The importance of the ozone layer and how CFCs react with ozone.
- How hardness occurs in tap water, its effects and how to remove it.
- Oils and fats, testing for healthier unsaturated fats, and making soap from vegetable oil.
- The composition and action of washing powder and detergents.

Electrolysis

Some compounds **conduct electricity** when they are **molten** (melted) or in **solution**, but not otherwise. In these cases, the liquid or solution must contain **ions**. A liquid or solution that conducts electricity is called an **electrolyte**. An electrolyte can be separated into its constituent parts by **electrolysis**.

When a direct current is passed through an electrolyte, the compound will **decompose** (break down) and elements will be produced. This is because the ions move to the electrode of opposite charge:

- The ions that are positively charged move towards the negative electrode (the cathode); they are called **cations**, e.g. H^+, Cu^{2+}, Al^{3+}.
- The ions that are negatively charged move towards the positive electrode (the anode); they are called **anions**, e.g. Cl^-, OH^-, SO_4^{2-}.

When they get there, they lose their charges, i.e. they are **discharged**. The negative ions lose electrons to the anode, and the positive ions gain electrons from the cathode, to form elements.

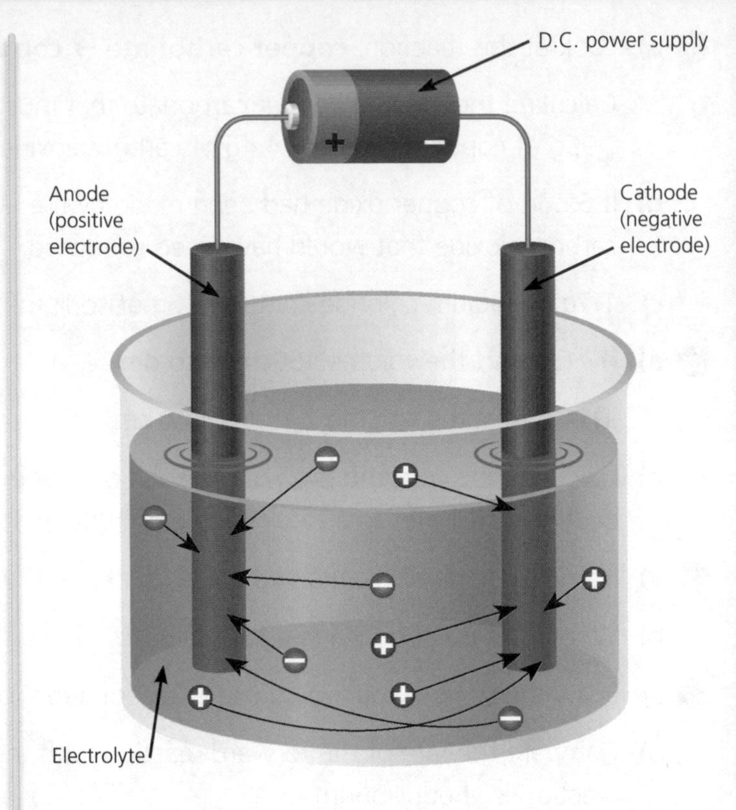

The electrolysis of copper(II) sulfate solution is carried out using this apparatus. Bubbles are seen at the carbon anode as oxygen is formed, and copper metal forms on the carbon cathode. The blue colour of the solution will slowly fade as the copper deposits on the cathode.

HT The reactions at the electrodes can be written as half-equations. This means that separate equations are written to show what is happening at each of the electrodes during electrolysis.

The electrolysis of copper(II) sulfate solution can be represented by the following two half-equations:

- At the cathode:

$$Cu^{2+} + 2e^- \longrightarrow Cu$$

- At the anode:

$$4OH^- - 4e^- \longrightarrow O_2 + 2H_2O$$

HT The electrolytes sulfuric acid and sodium hydroxide both produce hydrogen at the cathode and oxygen at the anode during electrolysis. It takes less energy to make hydrogen at the cathode rather than sodium when sodium hydroxide undergoes electrolysis.

The reactions that happen at the electrodes are:

- At the cathode:

$$2H^{+} + 2e^{-} \xrightarrow{\text{Reduction}} H_2$$

- At the anode:

$$4OH^{-} - 4e^{-} \xrightarrow{\text{Oxidation}} O_2 + 2H_2O$$

Electrolysis of Sulfuric Acid

When dilute sulfuric acid undergoes electrolysis:
- the hydrogen cations are attracted to the cathode and form hydrogen gas
- the hydroxide anions are attracted to the anode and form oxygen gas.

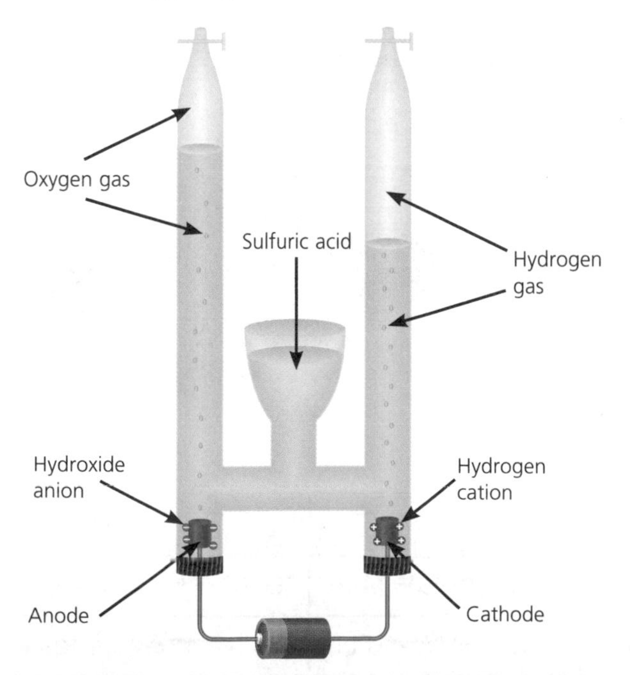

The electrolysis of sodium hydroxide solution in the above apparatus will also produce hydrogen at the cathode and oxygen at the anode.

Testing the Products

The gases made during electrolysis can be tested as follows:
- Hydrogen burns with a squeaky pop when tested with a lighted splint.

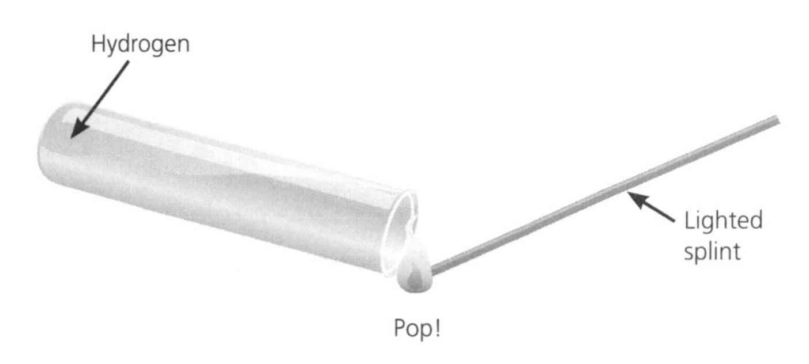

- Oxygen re-lights a glowing splint.

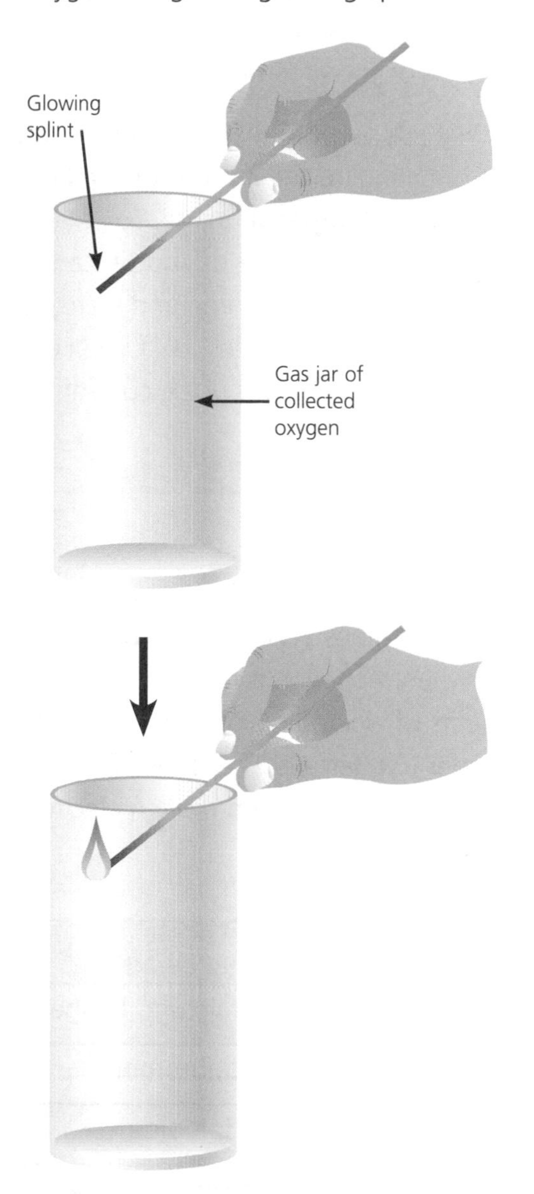

Controlling the Amount of Substance Produced

The amount of substance produced during electrolysis is determined by the size of the current and the length of time it is flowing for.

More substance is made if:

- a larger current flows
- the current flows for a longer time.

HT Calculating the Amount of Substance Produced

The amount of substance made in electrolysis is proportional to the current and the time that it flows:

- The amount of substance made increases if the current or the time increases.
- The amount of substance made decreases if the current or the time decreases.

Because they are proportional, you can calculate the effect of changing the current or the time on the amount of substance made. For example, to make 5 times the amount of substance you could keep the current the same and increase the time it flows for by 5 times, or if you keep the time the same you could use a current 5 times as large.

Example 1

4g of copper was made electrolysing copper(II) sulfate solution. How much copper would be made in the same time by increasing the current from 2 amps to 6 amps?

The current has increased by $\frac{6}{2}$ = 3 times. So the mass of copper made will increase by 3 times; 3 × 4g = **12g**.

Example 2

2.5 dm^3 of hydrogen gas was made during the electrolysis of sulfuric acid solution in 10 minutes. If the current remained the same, how long would it take to produce 1.0 dm^3 of hydrogen?

The volume of hydrogen made has reduced by $\frac{1.0}{2.5}$ = 0.4 times. So the time needed is 10 min × 0.4 = **4 minutes**.

Molten Electrolytes

Ions are in fixed positions in ionic solids but they are free to move when molten. It is much easier to work out the products of electrolysis of liquids because there is no water to interfere. The table below shows the elements made when certain molten electrolytes undergo electrolysis:

Liquids	Elements Made	At the Cathode	At the Anode
Aluminium oxide, $Al_2O_3(l)$	Aluminium, Oxygen	HT $Al^{3+}(l) + 3e^- \longrightarrow Al(l)$	$2O^{2-}(l) \longrightarrow O_2(g) + 4e^-$
Lead bromide, $PbBr_2(l)$	Lead, Bromine	HT $Pb^{2+}(l) + 2e^- \longrightarrow Pb(l)$	$2Br^-(l) \longrightarrow Br_2(g) + 2e^-$
Lead iodide, $PbI_2(l)$	Lead, Iodine	HT $Pb^{2+}(l) + 2e^- \longrightarrow Pb(l)$	$2I^-(l) \longrightarrow I_2(g) + 2e^-$
Sodium chloride, $NaCl(l)$	Sodium, Chlorine	HT $Na^+(l) + e^- \longrightarrow Na(l)$	$2Cl^-(l) \longrightarrow Cl_2(g) + 2e^-$

Fuel Cells

The reaction between hydrogen and oxygen is an exothermic reaction. This equation shows the reaction in a fuel cell.

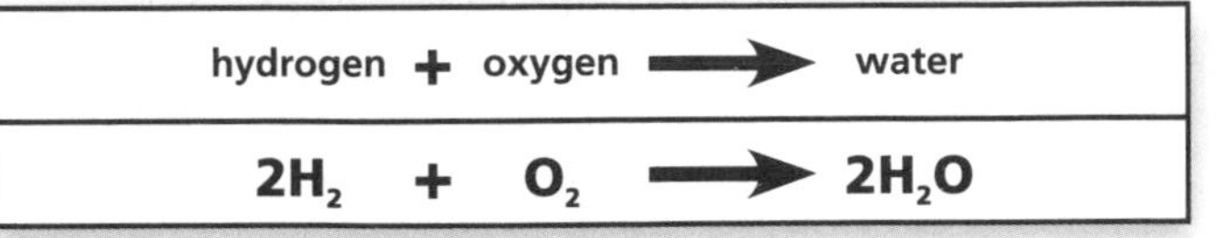

hydrogen + oxygen ⟶ water
$2H_2 + O_2 \longrightarrow 2H_2O$

Hydrogen reacts with oxygen in a **fuel cell** to produce an **electric current**. The energy from the reaction is used to create a **potential difference** (**pd**). A fuel cell is very efficient at producing electrical energy.

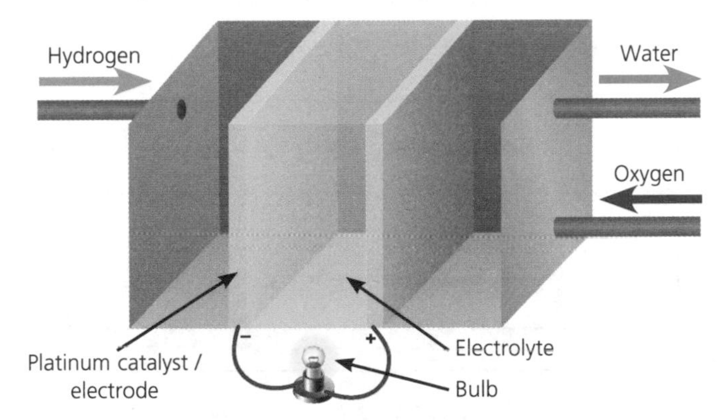

Fuel cells can be used to provide electrical power in spacecraft. An advantage of this is that the water made is pollution-free and can be used as drinking water for the crew.

The car manufacturing industry is very interested in developing fuel cells as a possible pollution-free method of powering electric cars.

Most cars use fossil fuels. These are non-renewable and emit large amounts of carbon dioxide, which is linked with global warming and climate change.

Fuel cells do not produce carbon dioxide and the hydrogen used in fuel cells is in plentiful supply from the decomposition of water.

HT In a fuel cell, each hydrogen molecule loses two electrons at the anode (the positive electrode) to form two hydrogen ions. This is **oxidation**: $H_2 - 2e^- \longrightarrow 2H^+$.

The hydrogen ions then move through the electrolyte towards the cathode (the negative electrode), and the electrons travel around the circuit.

Each oxygen molecule gains four electrons at the cathode. This is **reduction**: $O_2 + 4e^- \longrightarrow 2O^{2-}$.

The oxygen ions and hydrogen ions combine to form water. This is known as a **redox reaction** because **red**uction and **ox**idation both happen.

Some of the benefits of using a fuel cell are that they do not pollute, they are very efficient, they transfer energy directly, they have few stages and are simple to construct. Fuel cells are also lightweight, compact and have no moving parts.

There may be an environmental cost in using fuel cells – the hydrogen and oxygen may be produced using energy created by burning fossil fuels, and the catalysts used inside a fuel cell are often poisonous and will need careful disposal when the fuel cell has come to the end of its useful life.

The energy level diagram shown below is for the reaction between hydrogen and oxygen:

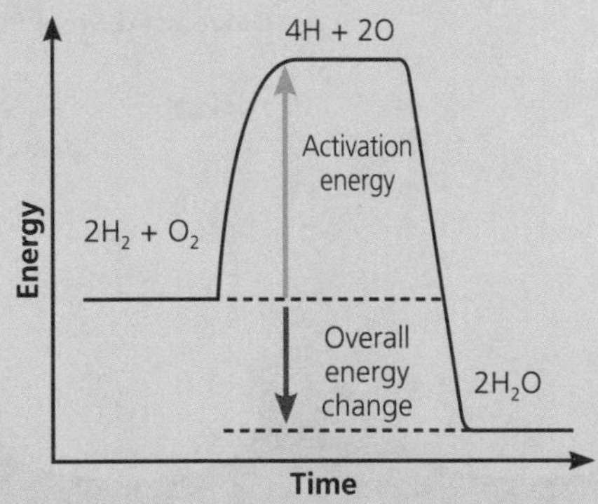

The hydrogen and oxygen molecules start at a certain energy level. Before they can react, some bonds have to be broken, and to do this the energy level has to be increased **(activation energy)**. The oxygen and hydrogen then bond together. So much energy is given out that the product has less energy than the reactants. This is an **exothermic reaction**.

You may be asked to draw an energy level diagram for a similar reaction if you are given information about the overall energy change.

Rusting

Rust is a form of hydrated iron(III) oxide that forms when iron or steel combines with both oxygen (or air) and water in a **redox reaction**.

iron + oxygen + water → hydrated iron(III) oxide

Iron and steel can be **protected** from rusting by being coated in oil, grease or paint. This stops the water and air coming into contact with the metal.

Other methods of protecting iron and steel from rusting include:
- galvanising
- alloying
- tin plating
- sacrificial corrosion.

Galvanising coats the iron or steel with zinc. The zinc stops the iron or steel from coming into contact with air and water, and it corrodes before the iron, acting as a sacrificial metal.

HT **Tin plating** coats the iron or steel with a layer of tin, which acts as a barrier to air and water. However, when the tin plating is scratched, the iron will corrode losing electrons.

Sacrificial protection works by having a more reactive metal, for example magnesium or zinc, in contact with the iron or steel. The metal will corrode, losing electrons in preference to the iron and thus protecting the iron.

HT

Redox Reactions

Rusting is a redox reaction because:
- iron loses electrons (oxidation)
- oxygen gains electrons (reduction).

A reagent that removes electrons from another substance is called an **oxidising agent**.

A reagent that gives electrons to another substance is called a **reducing agent**.

Some changes can be made using an oxidising agent. The following reactions are examples of oxidation because electrons are lost:

$$Fe \rightarrow Fe^{2+} + 2e^-$$

$$Fe^{2+} \rightarrow Fe^{3+} + e^-$$

$$2Cl^- \rightarrow Cl_2 + 2e^-$$

The reverse reactions can be done using a reducing agent. They are reduction reactions because electrons are gained:

$$Fe^{2+} + 2e^- \rightarrow Fe$$

$$Fe^{3+} + e^- \rightarrow Fe^{2+}$$

$$Cl_2 + 2e^- \rightarrow 2Cl^-$$

For other examples of redox reactions with electron transfer, look back at all the reactions at electrodes during electrolysis.

Displacement Reactions

Some metals are more reactive than other metals, so they will **displace** less reactive metals when they are combined.

For example, an experiment was carried out to see whether or not metals reacted with metal salt solutions.

A colour change or an increase or decrease in the temperature of the reaction mixture were used to indicate that a reaction had taken place. The results are shown in the table below:

	Zinc	Tin	Iron	Magnesium
Zinc chloride	✗	✗	✗	✓
Tin(II) chloride	✓	✗	✓	✓
Iron(II) chloride	✓	✗	✗	✓
Magnesium chloride	✗	✗	✗	✗

✗ = No reaction, ✓ = Reaction

The results of this experiment can be used to determine the **order of reactivity** of the four metals:

- Magnesium displaced the other three metals and so is the most reactive.
- Tin did not displace any and so is least reactive.
- Zinc displaced iron and tin, so zinc is more reactive than iron and tin.

Therefore, the reactivity order (from most to least reactive) is magnesium, zinc, iron, tin.

If you know the order of reactivity, you can use it to make predictions about whether one metal will displace another.

HT You should be able to construct formula equations for displacement reactions such as:

$$Mg(s) + ZnCl_2(aq) \rightarrow Zn(s) + MgCl_2(aq)$$

$$Zn(s) + Pb(NO_3)_2(aq) \rightarrow Zn(NO_3)_2(aq) + Pb(s)$$

$$Zn(s) + SnCl_2(aq) \rightarrow ZnCl_2(aq) + Sn(s)$$

Displacement reactions are redox reactions. In the third example above:

- Zinc atoms lose electrons to become zinc ions – oxidation.
- Tin ions gain electrons to become tin metal – reduction.

Displacement Reaction

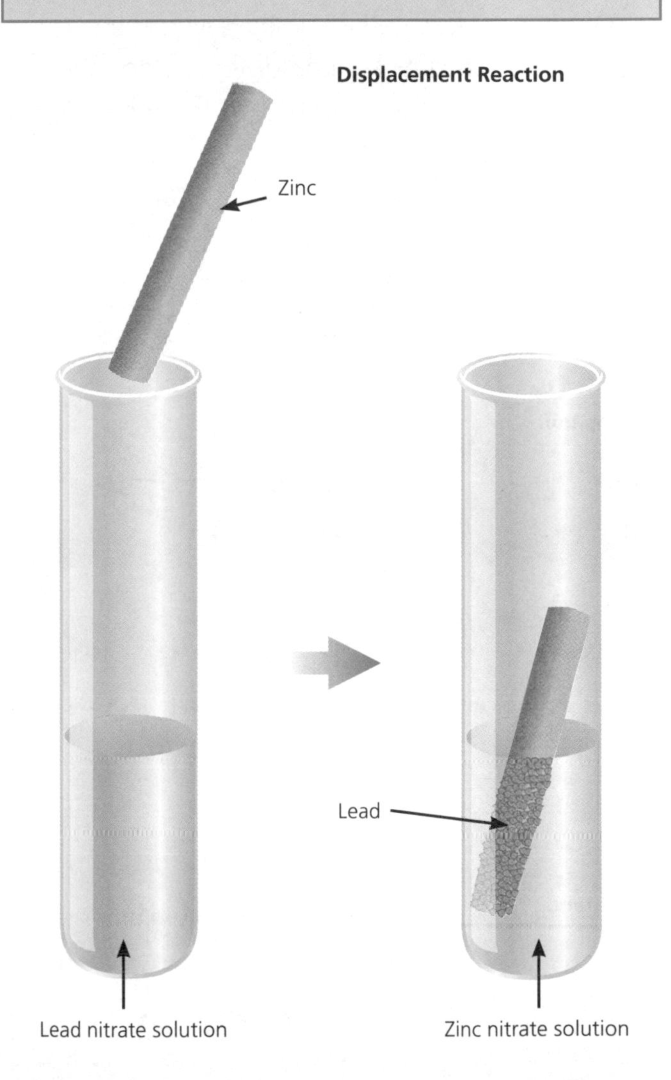

Ethanol and Alcohols

Ethanol (C_2H_5OH) is an alcohol. Its displayed formula is shown below:

```
    H  H
    |  |
H - C - C - O - H
    |  |
    H  H
```

Ethanol contains carbon and hydrogen, but because it also contains oxygen it is **not** a hydrocarbon. Ethanol has many uses, for example it can be used to make alcoholic drinks, to make solvents such as methylated spirits, and as fuel for cars.

HT Alcohols have the general formula $C_nH_{(2n+1)}OH$. For example, propanol is C_3H_7OH and pentanol is $C_5H_{11}OH$.

The displayed formulae of alcohols all have an **–OH** group like in ethanol. However, the size of the carbon chain will be different, and the –OH group can be anywhere on the molecule. For example:

Methanol

```
    H
    |
H - C - O - H
    |
    H
```

Propanol

```
    H   H   H
    |   |   |
H - C - C - C - O - H
    |   |   |
    H   H   H
```

Butanol

```
    H   H   H   H
    |   |   |   |
H - C - C - C - C - O - H
    |   |   |   |
    H   H   H   H
        |
        H
```

Pentanol

```
    H   H   H   H   H
    |   |   |   |   |
H - C - C - C - C - C - O - H
    |   |   |   |   |
    H   H   H   H   H
```

Making Ethanol

Ethanol can be made by **fermentation**. Yeast is used to ferment glucose solution (i.e. a sugar dissolved in water):

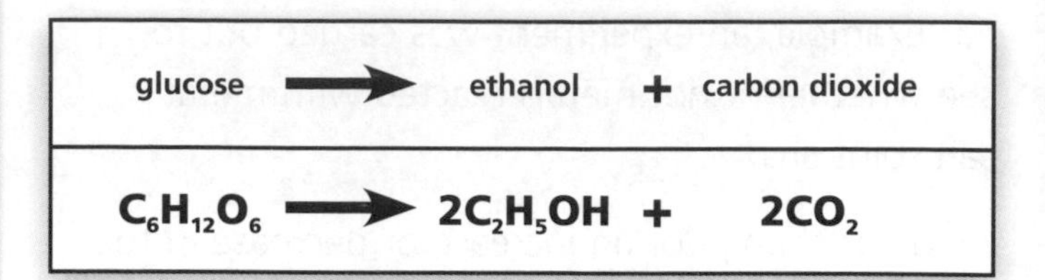

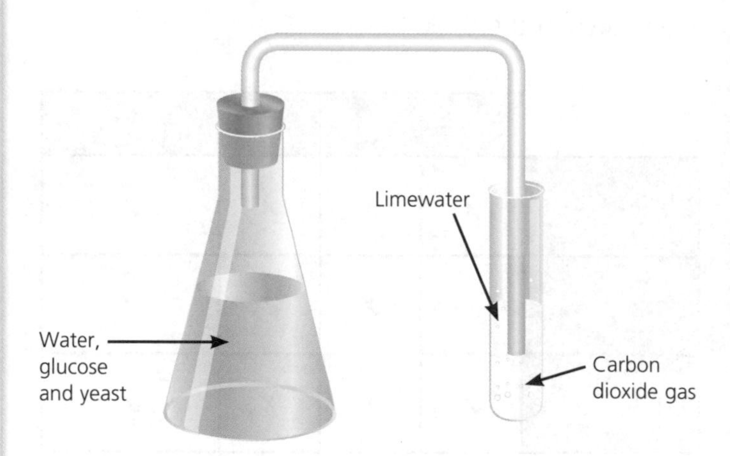

The apparatus used in fermentation prevents air (oxygen) reaching the fermentation mixture.

The fermentation mixture has to be kept between 25°C and 50°C for a few days. This is the **optimum temperature** for the enzymes in the yeast to convert the glucose into ethanol.

HT The absence of air from fermentation prevents the formation of ethanoic acid by oxidation of the ethanol.

The temperature of the fermentation mixture has to be kept between 25°C and 50°C (the optimum temperature) because:

- if it falls below the optimum temperature, the yeast becomes inactive
- if it rises above the optimum temperature, the enzymes in the yeast denature and stop working.

Pure ethanol can be extracted from a fermentation mixture by **distillation**:

Distillation Equipment

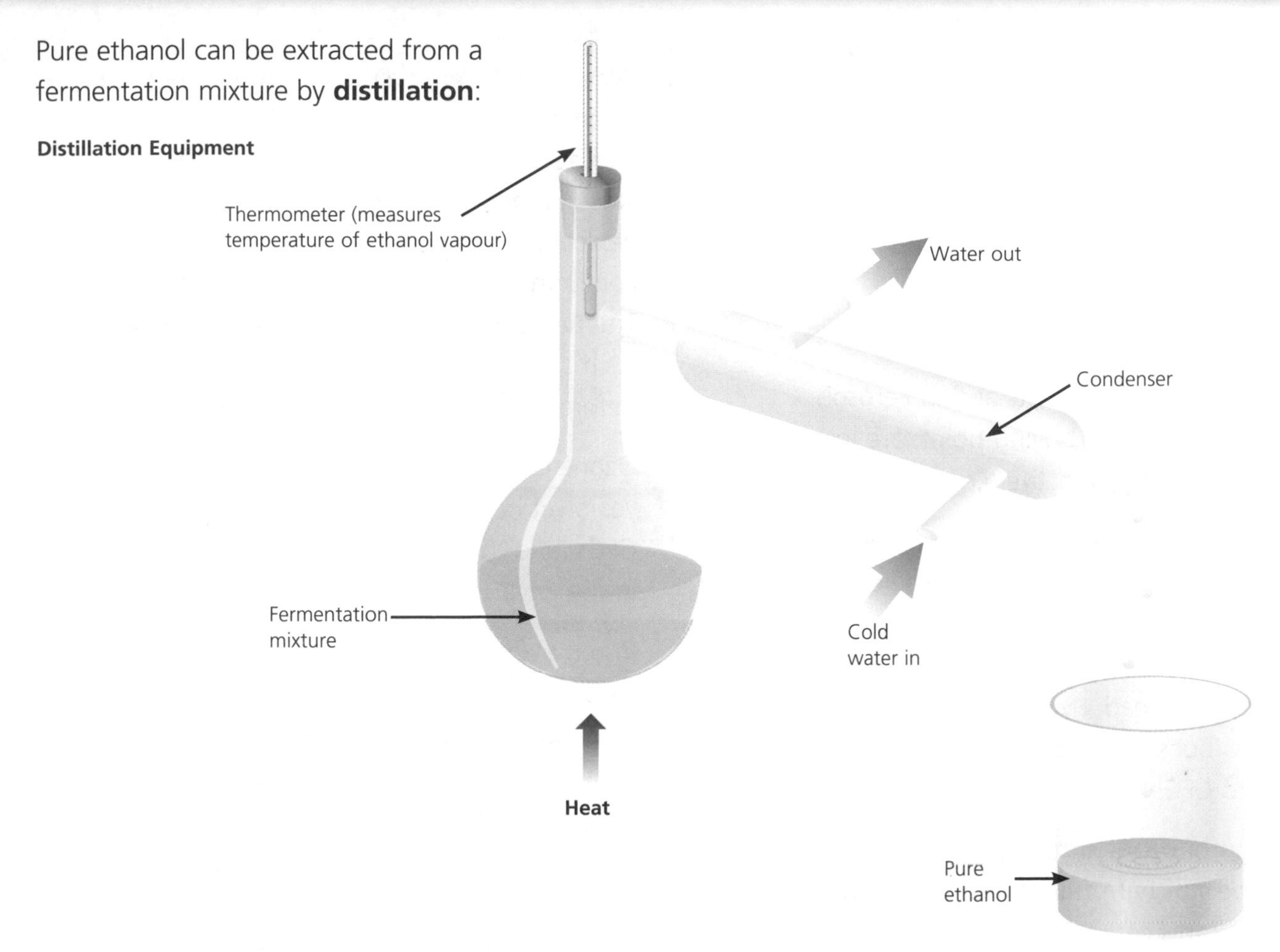

Ethanol made by fermentation is a renewable fuel; the raw material, sugar, will not run out as it is obtained from plants.

In another method, ethene can be **hydrated** to make ethanol by passing it over a heated phosphoric acid catalyst with steam.

Ethanol is made this way for industrial use only.

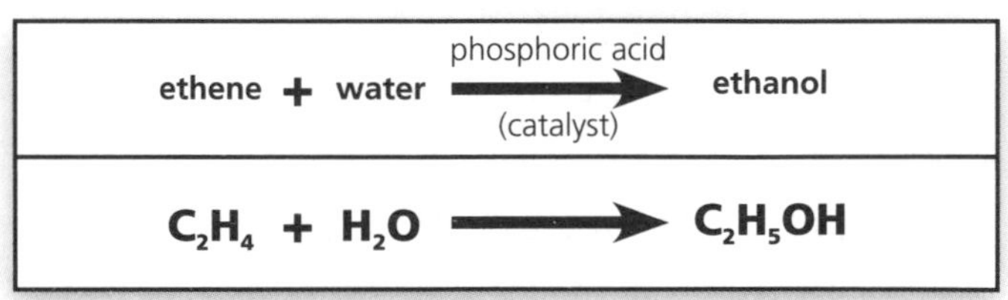

Ethanol made by hydration of ethene is a non-renewable fuel because ethene is obtained from the cracking of crude oil, which is a finite resource.

Comparing the Methods of Making Ethanol

Fermentation uses fairly mild conditions, atmospheric pressure and a temperature of 40°C. It is a good example of a batch process. The raw materials are extracted from plants so it is a sustainable process. The drawbacks are that it takes a long time, the ethanol has to go through a purification process, which uses lots of energy, and the yield is low as the yeast stops working when the ethanol level gets too high. If the carbon dioxide also made is not used the atom economy is less than 100%.

The **hydration** of ethene needs a high temperature and a high pressure but this has the advantages of being a continuous process and the product needing very little purification. It has a high yield and the atom economy is 100%. But this is not a sustainable process because the ethene is made from non-renewable crude oil.

Depletion of the Ozone Layer

Ozone (**O_3**) is a form of oxygen found in a layer high up in the atmosphere (the stratosphere). Ozone filters out and stops harmful ultraviolet (UV) light from reaching the surface of the Earth.

Chlorofluorocarbons (**CFCs**) are organic molecules that contain chlorine, fluorine and carbon.

CFCs were used as refrigerants and in aerosols because they have a low boiling point, are insoluble in water and are very unreactive (i.e. chemically inert).

Scientists argued that CFCs were responsible for the destruction of ozone in the atmosphere. The public has accepted this and the use of CFCs in the UK is now banned to stop any more damage to the ozone layer. Instead, hydrocarbons (alkanes) or hydrofluorocarbons (HFCs) are now used as safer alternatives to CFCs.

The depletion of ozone in the atmosphere allows increased levels of harmful ultraviolet light to reach the Earth, and this can:

- speed up the risk of sunburn
- cause increased ageing of the skin
- cause skin cancers
- lead to increased risk of cataracts.

When a CFC molecule in the stratosphere is hit by ultraviolet light, a very reactive chlorine atom is produced. A chlorine atom is called a chlorine **radical**.

$CFCl_3$	$\longrightarrow$	$CFCl_2$	+	Cl
(CFC)				(Chlorine radical)

Chlorine radicals react with ozone molecules, but once an ozone molecule is destroyed, the chlorine radical is remade and can go on to destroy many more ozone molecules.

CFCs take a very long time to be removed from the stratosphere because they are so unreactive.

HT When CFCs were first made they were seen to be very useful, particularly because of their inertness. However, the enthusiasm for their use was dampened when the link between ozone depletion and CFCs was found. The international community decided to ban the use of CFCs. The ban started in the developed countries, such as the UK. However, it was a few years before less developed countries followed suit.

Ozone depletion will continue for up to 100 years yet because some countries are still using CFCs, and their inertness causes them to stay in the environment for a long time.

When ozone absorbs ultraviolet light in the stratosphere, the energy in the light causes a **covalent bond** in the ozone molecule to break. The ozone molecule is split into normal oxygen (O_2) and an oxygen atom. When this happens, the covalent bond can be broken evenly so that each atom retains one electron to form radicals, or unevenly to make **ions**.

The following reactions occur when a chlorine radical attacks an ozone molecule:

- The chlorine radical reacts with an ozone molecule to form a chlorine monoxide molecule and an oxygen molecule:

$$Cl + O_3 \longrightarrow ClO + O_2$$

- The chlorine monoxide molecule then reacts with an oxygen atom to produce a chlorine radical and an oxygen molecule:

$$ClO + O \longrightarrow Cl + O_2$$

The chlorine radical is **regenerated** in this **chain reaction** and can go on to destroy many more ozone molecules. Therefore, a few chlorine atoms can destroy large amounts of ozone.

Hardness of Water

Water can be described as being hard or soft:

- **Soft** water lathers well with soap.
- **Hard** water does not lather with soap.

Both hard and soft water will lather with a soapless detergent.

Calcium ions and magnesium ions from dissolved salts cause hardness in water. Rainwater must come into contact with rocks that contain calcium ions or magnesium ions to become hard. The rocks in soft water areas do not contain these metal ions. There are two types of hardness in water:

- **Permanent hardness** is caused by dissolved calcium sulfate. It cannot be destroyed by boiling.
- **Temporary hardness** forms when rainwater comes into contact with rock that contains calcium carbonate, e.g. chalk, marble or limestone. It can be removed by boiling.

Rainwater contains dissolved carbon dioxide, which makes it slightly acidic. When rainwater reacts with rock containing calcium carbonate it makes soluble calcium hydrogencarbonate:

calcium carbonate + water + carbon dioxide → calcium hydrogen-carbonate

When water is boiled, calcium hydrogencarbonate decomposes to form insoluble calcium carbonate, water and carbon dioxide, i.e. the hardness is removed from the water:

calcium hydrogen-carbonate → calcium carbonate + water + carbon dioxide

HT $$Ca(HCO_3)_2(aq) \rightarrow CaCO_3(s) + H_2O(l) + CO_2(g)$$

Removing Hardness from Water

All types of hardness can be **removed** from water by:

- adding washing soda (sodium carbonate crystals)
- passing the water through an **ion-exchange column**.

When hard water is passed through an ion-exchange column, the calcium ions and magnesium ions attach to the resin inside the column, and sodium ions are released into the water, removing the hardness.

Ion-exchange Column

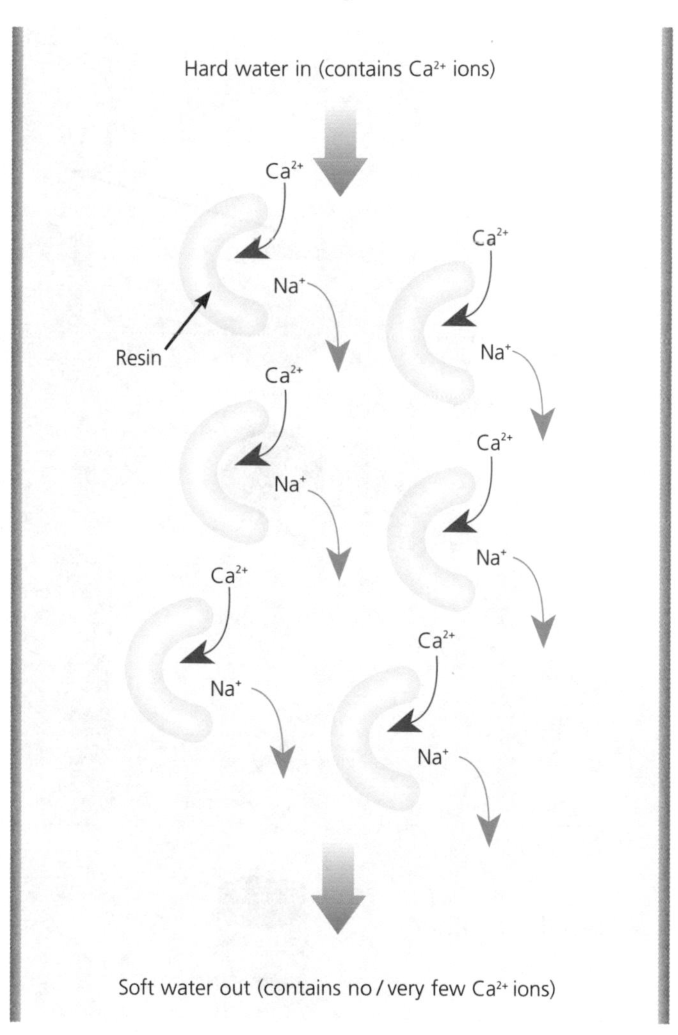

HT When washing soda (sodium carbonate crystals) is used to soften water, the calcium ions and magnesium ions are removed from the water as they form precipitates of insoluble calcium carbonate and magnesium carbonate.

Hardness of Water

Measuring Water Hardness

Hardness in water can be measured by adding soap solution to some water until a permanent lather is produced after it is shaken for five seconds.

The number of drops of soap solution added can be counted, or a burette can be used to measure the volume of soap solution added.

1. Add soap solution to the water

2. Shake for five seconds

3. Look for a lather

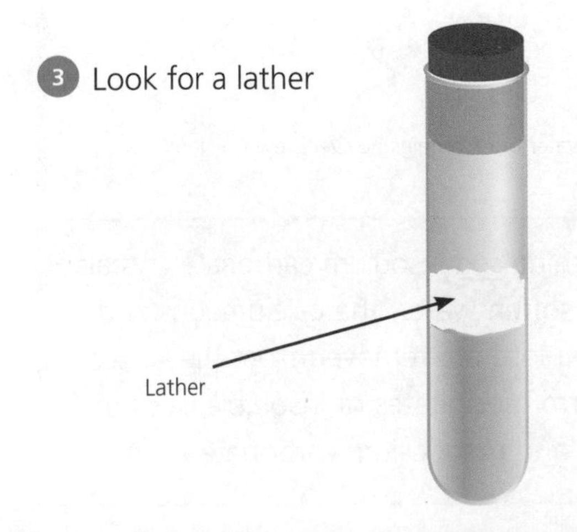

You may be asked to outline a procedure to test some samples of water based on this method. If you are comparing samples of water then you must use the same volume of water each time. You may have to test the samples before and after boiling (for the same amount of time) to identify temporary hardness.

In your exam, you will be expected to interpret data from an experiment to measure hardness in water, such as the information shown in this table:

	Drops of Soap Solution Required to Produce a Lather	
Water Sample	**Before Boiling**	**After Boiling**
A	14	14
B	2	2
C	12	2
D	14	8
E	15	2

From the results in the table, it can be seen that:

- Sample **A** is permanently hard water because there is no change on boiling.
- Sample **B** is soft water because it needs very little soap to form a lather.
- Sample **C** is temporarily hard as the hardness is removed by boiling.
- Sample **D** contains both permanent and temporary hardness because some of the hardness is removed by boiling but it still needs lots of soap before a lather forms.
- Sample **E** is the hardest of all the samples because it needs the most soap to form a lather.

Natural Fats and Oils

At room temperature, oils are liquids and fats are solids. Oils and fats are **esters** that can be obtained from animals or vegetables. Animal oils include lanolin, lard and tallow. Examples of oils from plants include olive oil, rapeseed oil and maize oil.

Oils and fats can be:
- **saturated**, i.e. all the carbon-carbon bonds are single bonds **or**
- **unsaturated**, i.e. the molecules contain at least one carbon-carbon double bond.

Bromine reacts with the carbon-carbon double bond in unsaturated fats and oils; this is an addition reaction and a colourless dibromo compound is formed and so:
- the bromine water will change from orange to colourless if the fat is unsaturated (see diagram 1)
- the bromine water will remain orange if the fat is saturated (see diagram 2).

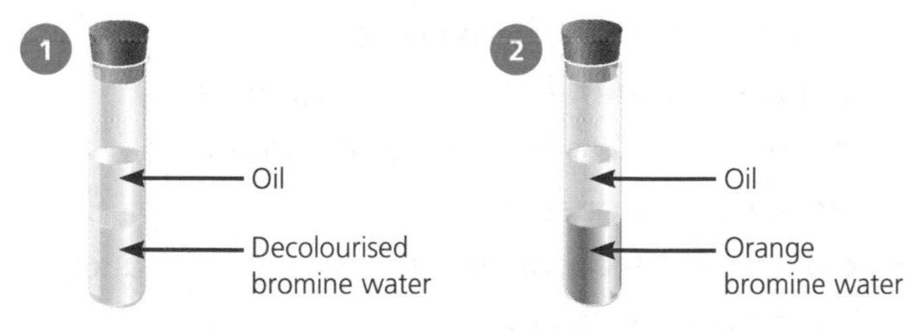

Many animal oils and fats are saturated. Many vegetable oils and fats are unsaturated.

It is better to have more unsaturated than saturated oils and fats in a diet in order to reduce the build up of cholesterol in the body. This is healthier for the heart because cholesterol builds up in the blood vessels, causing the heart to have to do more work. It can eventually lead to heart disease.

When testing oils and fats with bromine water:
- saturated oils and fats do not contain a carbon–carbon double bond and do not react, and so the bromine water stays orange
- the carbon–carbon double bonds in unsaturated oils and fats are able to react with the bromine water, turning the bromine water colourless.

Mixing Fats and Oils

Oil and water are **immiscible**, i.e. they do not normally mix. When a vegetable oil is added to water and shaken well, it makes an **emulsion**. The shaking breaks up the oil into small droplets that **disperse** (spread out) in the water. For example:
- milk is an oil-in-water emulsion that is mostly water with tiny droplets of oil dispersed in it
- butter is a water-in-oil emulsion that is mostly oil with droplets of water dispersed in it.

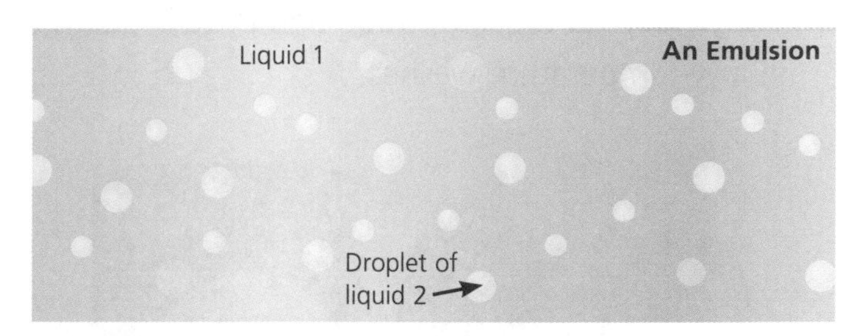

When a vegetable oil is reacted with hot sodium hydroxide, the sodium hydroxide splits up the oil or fat molecules into glycerol and soap (a sodium salt of a long-chain fatty acid). This process is called **saponification**.

Margarine is also made from vegetable oils (unsaturated), which are reacted with hydrogen using a nickel catalyst. The product is a solid saturated fat, which can then be blended with other materials to make it taste and look like butter.

Oils and fats are very important raw materials for the chemical industry. There is much interest in how vegetable oils can be converted into biodiesel to be used as a renewable replacement for diesel, which is obtained from crude oil.

Saponification is the breakdown of an oil or fat molecule by heating it with sodium hydroxide:

fat / oil + sodium hydroxide → soap + glycerol

This reaction can be described as **hydrolysis** – it involves breaking up ester groups in the oil molecule using an alkali.

Washing Powder

A washing powder is a **mixture**. The main components are:

- **detergent** – to do the cleaning
- **bleach** – to remove coloured stains
- **water softener** – to soften hard water (hard water contains soluble calcium and magnesium compounds that react with soap)
- **optical brightener** – to make whites appear brighter
- **enzymes** – to break up food and protein stains in low-temperature washes.

When clothes are washed, the water is the **solvent** (the liquid that does the dissolving), and the washing powder is the **solute** (the solid that gets dissolved) because it is soluble (i.e. it dissolves) in water. The resulting mixture of solvent and solute is a **solution**.

Low-temperature washes are used to wash delicate fabrics (i.e. ones that would shrink in a hotter wash or that have a dye that could run) and they do not denature enzymes in biological powders. An advantage of low-temperature washes is that they save energy.

Detergent molecules have two ends:

- a hydrophilic head that likes to be in water and helps the detergent dissolve.
- a hydrophobic tail that does not like water and helps the molecule attach to grease and oils.

Some stains will not dissolve in water; they are **insoluble**. Dry-cleaning solvents are used when a stain is insoluble in water. The clothes are still washed in a liquid but it does not contain water. Different solvents will dissolve different stains (see the table).

Stain	Solvent
Ball-point pen	Methylated spirits (ethanol)
Blood	Water
Shoe polish	White spirit
Coffee	Water
Correcting fluid	White spirit or nail varnish remover

You do not have to remember the details of this table but you may be asked to use similar information to choose which solvent to use to remove a stain.

Washing-up Liquid

Washing-up liquid contains:

- **detergent** – to do the cleaning
- **water** – to dissolve and dilute the detergent so that it pours out of the bottle
- **water softener** – to soften hard water
- **rinse agent** – to help the water drain off crockery so it dries quickly
- **colour and fragrance** – to make the product more attractive to buy.

In your exam you may be asked to interpret data about the effectiveness of washing-up liquids and washing powders. The data could be presented in a table or a graph. For example:

Washing-up Liquid	Plates washed per $10cm^3$	Cost per $10cm^3$
A	47	0.20
B	32	0.15
C	41	0.17

Can you pick out which washing-up liquid washed the most plates, or what the relationship is between cost and the number of plates washed? Is the cheapest washing-up liquid always value for money?

Detergent Molecules

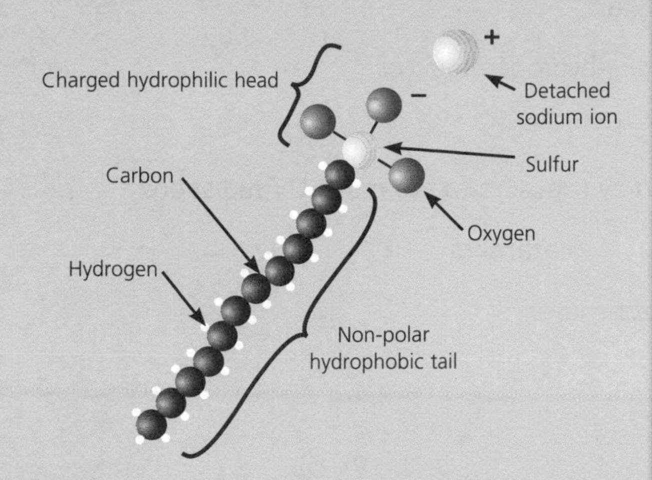

The charged 'head' is attracted to water molecules because of its polarity (electrical charge), so it is known as **hydrophilic** (water-loving).

The hydrocarbon tail is non-polar and so it is not attracted to water molecules. It is therefore known as **hydrophobic** (water-hating).

How Detergents Work

The diagram below shows how the detergents in washing-up liquid work:

1. The hydrophobic end of the detergent molecule forms strong intermolecular forces (attractive forces between the molecules) with the oil droplet, while the hydrophilic end of the molecule forms strong intermolecular forces with the water.
2. As more and more detergent molecules are absorbed into the oil droplet, the oil is eventually lifted off the plate.
3. When it is totally surrounded, the oil droplet can be washed away, leaving the plate clean.

Dry-cleaning Solvents

The molecules that make up a grease stain are attracted to each other by weak **intermolecular forces**. There are also weak intermolecular forces between the solvent molecules. The grease-stain substance will dissolve in a dry-cleaning solvent if the intermolecular forces are overcome. The new intermolecular forces between the grease-stain molecules and the solvent molecules are stronger than the ones that were present before. The solvent molecules then surround the grease-stain molecules.

The coating on photographic film is a protein and is broken down by the enzymes in washing powder. A student timed how long it took the washing powder to break up the coating. The results are shown in the table:

Water (cm^3)	Biological Powder (g)	Time Taken (s)
50	1	340
50	2	125
50	3	62
50	4	62

You could be asked to plot a graph to show how good the washing powder is at removing the protein 'stain' or you may be asked to decide on the best concentration of powder to use. You could predict the effect of repeating the experiment at 70°C or using a non-biological washing powder.

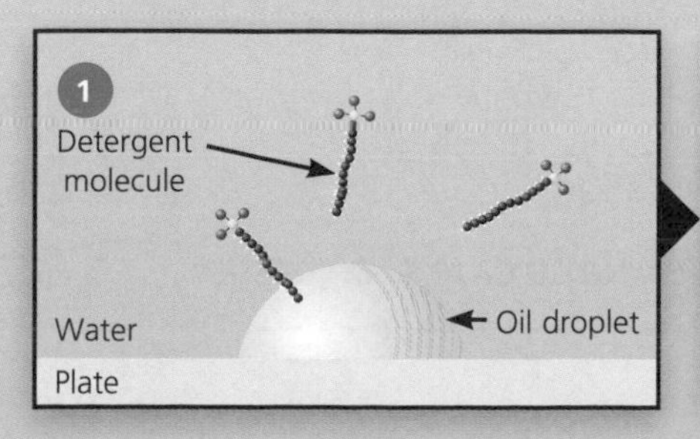

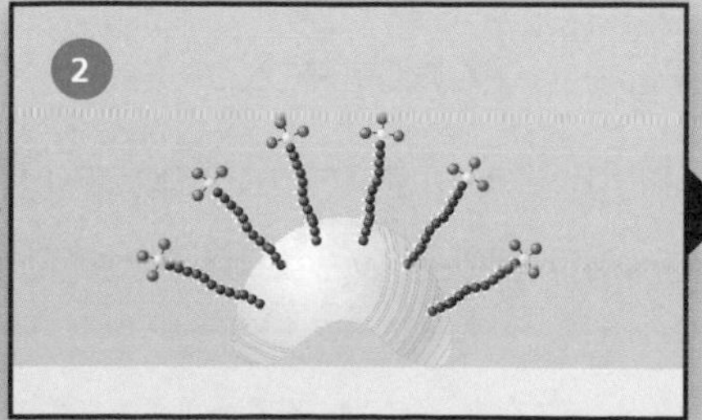

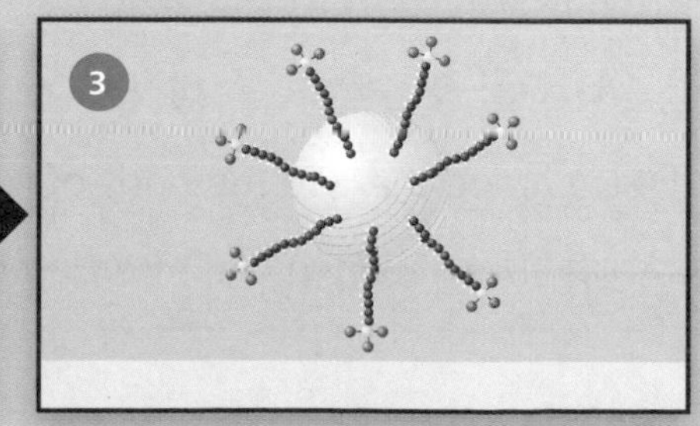

C6 Exam Practice Questions

1 a) Name the electrode that positive ions are attracted to. [1]

b) How do you change an ionic solid into an electrolyte? [1]

2 a) Write a word equation for the reaction in an oxygen–hydrogen fuel cell. [1]

b) Name two benefits of using fuel cells in a spacecraft. [2]

3 a) Select which of the following problems are caused by exposure to too much UV radiation: [1]

skin cancer **lung cancer** **sunburn** **cataracts**

b) Describe how CFCs are thought to damage the ozone layer. [3]

c) Describe how a radical is formed. [2]

4 a) Describe how permanent hard water can be made soft. [1]

b) Write a word equation to show how temporary hardness forms when rainwater comes into contact with limestone rocks. [1]

5 Outline the main difference between a saturated oil and an unsaturated oil. [2]

6 Matt and Joward made up four solutions of a biological washing powder of different concentrations. They put a small piece of photographic film in each one. The coating on the plastic film is a protein and is broken down by the enzymes in the washing powder. They timed how long it took the washing powder to break up the coating.

Water (cm^3)	Biological Powder (g)	Time to Break Up Stain (s)
50	1	390
50	2	105
50	3	62
50	4	62
50	5	62

a) Explain why they recommend using 3g of powder in $50cm^3$ of water as the best concentration. [1]

b) Calculate how much powder would be needed for a washing machine that used $2000cm^3$ water. [2]

HT c) Suggest how the results would be different if the experiment was carried out at 80°C. Give a reason for your answer. [3]

7 Noah makes 5.2g of copper by electrolysing copper(II) sulfate solution passing a current for 15 minutes. Calculate the mass of copper he would make by passing the same current for 2 hours. [2]

8 State which of the following reactions is oxidation and explain your choice.

A) $Fe^{2+} \rightarrow Fe^{3+}$ **B)** $Fe^{3+} \rightarrow Fe^{2+}$ **C)** $Cl_2 \rightarrow 2Cl^-$ [2]

9 Compare the two methods of making ethanol (fermentation and hydration). [6]

The quality of your written communication will be assessed in your answer to this question.

C1 Carbon Chemistry

1. **a)** Coal; Oil; Natural gas.
   b) Diesel comes from crude oil; This takes millions of years to make.
   c) **This is a model answer that demonstrates QWC and would score the full 6 marks:** The terminal would bring investment and jobs to the area improving the living standards of the people who live there. The jobs would be in the building industry and in companies that provide supplies and services to the visiting ships. However, if there is an accident and crude oil spills into the sea then it will float on the water and will damage the environment. Crude oil is poisonous to wildlife; it will coat the feathers of sea birds stopping them from flying or floating, and may even kill them. The crude oil will float ashore and damage the beaches and shore line.
   d) The crude oil is boiled; The fractions can be separated and collected because different hydrocarbons boil at different temperatures.
   e) It uses heat; and a catalyst; to break up large hydrocarbon molecules; that cannot be used in small hydrocarbon molecules; such as petrol **[Any three for 3 marks]**
   f) **i)** D **ii)** B
   g) Petrol

2. **a)** All the fuel is burned in a blue flame (complete combustion), all the energy is released; Only some of the fuel is burned in a yellow flame so less energy is released.
   b) Non toxic; Easy to control; No pollution; High energy value; Available; No need for storage **[Any two for 2 marks]**
   c) Methane + oxygen ➔ carbon dioxide + water

3. **a)** It is made by incomplete combustion of the fuel; It is toxic / poisonous.
   b) Increases oxygen; Reduces carbon dioxide.
   c) Damages trees; Erodes stonework; Corrodes metals; Kills plants; Kills fish. **[Any three for 3 marks]**

4. **a)** It does not rot.
   b) Waterproof; Breathable.

5. **a)** Ethanol; Ethanoic acid. **[Any for 1 mark]**
   b) The pigments used.

6. Small molecules have weaker forces of attraction between the molecules than large molecules; So less energy is needed to separate them.

7. $CH_4(g) + 2O_2(g) \rightarrow CO_2(g) + 2H_2O(l)$ **[All correct for 3 marks]**

8. There is less photosynthesis; Less carbon dioxide removed from air; Less oxygen made; The burning of the trees puts carbon dioxide into the air; Removes oxygen.

9. **a)** $2NaHCO_3 \rightarrow Na_2CO_3 + H_2O + CO_2$ **[1 mark for correct formula, 1 mark for correct balancing]**
   b) It changes shape.

C2 Chemical Resources

1. **a)** The soil is very fertile.
   b) The crust and the outer part of the mantle.
   c) It fits a wide range of evidence; It has been tested by many scientists.

2. **a)** Noise; Dust; Increased road traffic.
   b) $CaCO_3 \rightarrow CaO + CO_2$

3. **a)** It is harder; It is stronger.
   b) Tin
   c) It is hard; It is strong; It has a low melting point; It is gas proof. **[Any two for 2 marks]**

4. **a)** 550°C
   b) 34%
   c) The yield increases.

5. It contains the essential elements N (nitrogen); and P (phosphorus); It is soluble in water and so can be taken in by plant roots.

6. **a)** Hydrogen; Chlorine; Sodium hydroxide.
   b) Chlorine and sodium hydroxide.
   c) $2Cl^- - 2e^- \rightarrow Cl_2$ **[1 mark for correct formulae, 1 mark for correct balancing]**

7. The more dense oceanic plate is pushed under the continental plate; Down into the mantle where it melts; The result is a mountain range and possibly volcanoes.

8. Limestone is sedimentary rock with grains loosely stuck together – it is the softest; Marble is metamorphic, made from limestone that has been heated and pressured to make it more crystalline; Granite is igneous rock that is crystalline and the hardest of the three.

9. **This is a model answer that demonstrates QWC and would score the full 6 marks:** The Haber process uses an iron catalyst, which speeds up the rate of reaction so that the ammonia is made as fast as possible. The temperature used is 450°C. This is a compromise because you get a better yield of ammonia if you use a lower temperature, but the ammonia is then made very slowly. At a higher temperature the ammonia would be made faster but the yield would be very low. The pressure used is 200 atmospheres because this gives a fast reaction with a good yield without costing too much to achieve.

10. It is called eutrophication; The fertiliser gets into the water and causes algal blooms; This blocks out sunlight; Other plants die; Aerobic bacteria that rot the plant material use up all the oxygen; and most other living things in the water die. **[Any three for 3 marks]**

11. $2HNO_3 + Na_2CO_3 \rightarrow 2NaNO_3 + H_2O + CO_2$ **[1 mark for correct formulae, 1 mark for correct balancing]**

C3 Chemical Economics

1. **a)** They must collide; With sufficient energy.
   b) There are more particles in the same volume/ they are more crowded together.
   c) i) B **ii)** C
   d) A collision must have enough energy for the particles to react; If they do not have sufficient energy they will not react.

2. **a)** It is a substance that speeds up a chemical reaction; Without being used up.
   b) This is a model answer that demonstrates QWC and would score the full 6 marks: The main precaution taken in a flour mill is to prevent the flour dust mixing with the air. The flour dust will burn and any type of spark will set it off so all the electrical equipment has to be made so that it does not produce a spark. Flour dust has a very large surface area and when it mixes with air there is a very large chance of collision between the particles. When the dust–air mixture is set alight there is a very fast reaction – an explosion.

3. **a)** 106 **b)** 46
   c) i) 12 + 32 = 44g
   ii) 32 ÷ 12 = 2.6666g = 2.7g **[1 mark for calculation, 1 mark for correct answer to 2 significant figures]**

4. **a)** Carbon
   b) It is very hard; It has a high melting point.

5. **a)** 26°C
   b) Energy = 100 × 4.2 × 26 = 10 920J **[1 mark for calculation, 1 mark for correct answer]**

6. Most of the bonds in graphite are strong; Carbon atoms arranged in layers but some of the bonds between layers are weak; The electrons become delocalised (able to move around).

7. **a)** It takes a long time to make and test all the different versions of the drug; This work is carried out by highly qualified scientists; Large wage bill. **[Any two for 2 marks]**
   b) It must be tested; It must be shown to work; There must be trials on humans; It must be safe. **[Any two for 2 marks]**

C4 The Periodic Table

1. **a)** Atoms of the same element; With different numbers of neutrons.
   b) 2.8.4

2. **a) i)** Remove 1 electron from the atom.
   ii) Add 2 electrons to the atom.

b) Sodium atom **Sodium ion**

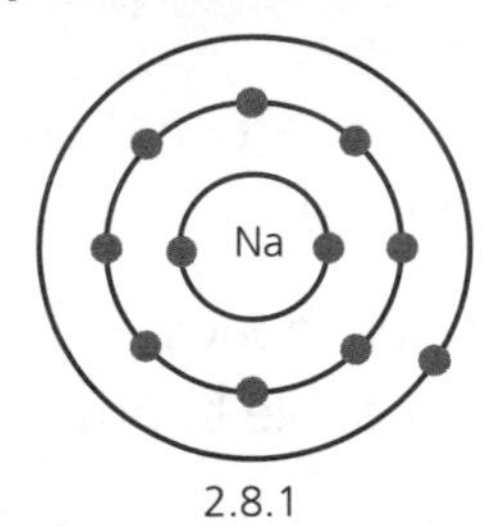

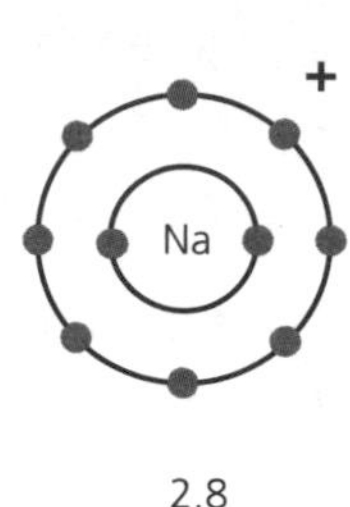

3. **a)** Chlorine
   b) Sodium iodide + chlorine ➔ sodium chloride + iodine
4. **a)** They are very strong.
   b) They are shiny.
   c) Non-toxic; Hard; Unreactive. **[Any one for 1 mark]**
5. It has equal numbers of protons and electrons.
6. **This is a model answer that demonstrates QWC and would score the full 6 marks:** There are strong forces holding the ions together in solid sodium chloride and a lot of energy is needed to break them apart. When sodium chloride is molten, the charged ions can move and carry the electrical charge. In methane there are no charged particles and all the electrons are involved in bonding so there is nothing to carry an electrical charge. The forces between methane molecules are weak and so only a small amount of energy is needed to separate the molecules.
7. **a) i)** $2NaI + Cl_2 \rightarrow 2NaCl + I_2$. **[1 mark for correct reactants, 1 mark for correct products]**
   ii) It is a displacement reaction.
   iii) it shows that chlorine is more reactive than iodine.
   b) $Br_2 + 2e^- \rightarrow 2Br^-$. **[1 mark for correct reactants, 1 mark for correct products]** The bromine gains electrons. **[1 mark]**
8. **a)** A material that conducts electricity with very little or no resistance.
   b) Benefit – No energy lost during power transmission; Super fast circuits; Powerful electromagnets. **[Any one for 1 mark]**
   Problem – They only work at very low temperatures; Expensive. **[1 mark]**
9. **a)** Calcium
   b) $Fe^{2+} + 2OH^- \rightarrow Fe(OH)_2$ **[1 mark for correct formula, 1 mark for correct balancing]**
10. **a)**

Mg^{2+} ion $[2.8]^{2+}$

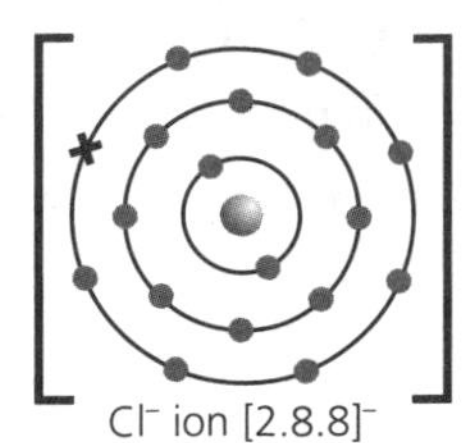

Cl^- ion $[2.8.8]^-$

   b) $MgCl_2$

11.

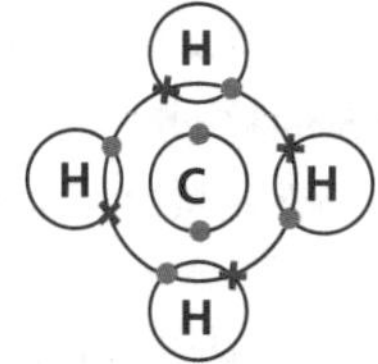

[1 mark for one shared pair of electrons, 1 mark for the rest of the diagram]

C5 How Much? (Quantitative Analysis

1. **a)** 7.95 + 4.4 = 12.35g **[1 mark for calculation, 1 mark for correct answer]**
   b) 63.6 ÷ 79.5 = 0.8, 0.8 × 44 = 35.2g **[1 mark for calculation, 1 mark for correct answer]**
   c) 117 – 46 = 71g **[1 mark for calculation, 1 mark for correct answer]**
2. **a) i)** 150 ÷ 1000 = $0.15dm^3$ **[1 mark for calculation, 1 mark for correct answer]**
   ii) 2.75 × 1000 = $2750cm^3$ **[1 mark for calculation, 1 mark for correct answer]**
   b) 10 ÷ 100 = 0.1, 0.1 × 2 = $0.2mol/dm^3$ **[1 mark for calculation, 1 mark for correct answer]**
3. **a)** Neutralisation
   b) The volume of acid or alkali added from a burette that just neutralises the test solution; Final volume – Start volume on a burette.
4. **a)** $\rightleftharpoons$
   b) The rate of the forward and reverse reactions are equal.
5. Sulfur; Air; Water; Catalyst is vanadium(v) oxide.
6. **a)** Citric acid
   b) Georgie.

7. **a) This is a model answer that demonstrates QWC and would score the full 6 marks:** Lead nitrate is added to the sodium chloride to obtain a precipitate of lead chloride. The mixture is then filtered (keeping the solution) and the lead chloride is left on the filter paper. This can be washed with distilled water and then put in an oven to dry at 50°C. The solution that passed through the funnel contains dissolved sodium nitrate. This is put in an evaporating basin and heated to evaporate the water, leaving behind the solid salt.
   b) Lead nitrate + sodium chloride ➔ lead chloride + sodium nitrate
   c) The sodium ions stay dissolved in the water, they do not take part in the reaction.
8. **a)** 16 ÷ 40 = 0.4 moles **[1 mark for calculation, 1 mark for correct answer]**
   b) 3 × 16 = 48g **[1 mark for calculation, 1 mark for correct answer]**
9. The reaction would be too slow if it were colder; It is a compromise between rate and yield.
10. Number of moles = 4.8 ÷ 24 = 0.2 moles. **[1 mark for calculation, 1 mark for correct answer]**

C6 Chemistry Out There

1. **a)** Cathode / negative
   b) Melt the solid or dissolve it in water.
2. **a)** Hydrogen + oxygen ➔ water
   b) They are lightweight; Compact; Have no moving parts; Produce water for use on board the spacecraft. **[Any two for 2 marks]**
3. **a)** Skin cancer, sunburn and cataracts.
   b) UV light hits the CFC; This breaks a bond and produces a chlorine atom; Chlorine atoms are very reactive; They will react with ozone; One chlorine atom will destroy many ozone molecules. **[Any three for 3 marks]**
   c) A covalent bond breaks evenly so that each atom retains one electron; An atom with an unpaired electron is a radical.
4. **a)** Add sodium carbonate (crystals) or pass it through an ion-exchange resin.
   b) Calcium carbonate + water + carbon dioxide ➔ calcium hydrogencarbonate
5. Saturated oils contain only single bonds; Unsaturated oils also contain double bonds.
6. **a)** It was the quickest, adding more powder did not improve the time.
   b) 2000 ÷ 50 =40, 40 × 3g = 120g **[1 mark for calculation, 1 mark for correct answer]**
   c) The high temperature will denature the enzymes in the powder; So the protein will not be removed; All concentrations are likely to give the same long time.
7. 2 hours = 15 minutes × 8; So 8 times as much copper is made; 5.2g × 8 = 41.6g **[1 mark for calculation; 1 mark for correct answer]**
8. **A** is oxidation; It is the only one involving loss of electrons.
9. **This is a model answer that demonstrates QWC and would score the full 6 marks:** The fermentation process to make ethanol uses mild conditions. The temperature used is about 40°C and it is carried out at normal atmospheric pressure. It is a batch process and the raw materials are sustainable because they come from plants. The process takes a long time and the yield is low. The ethanol has to be extracted by distillation, which uses a lot of energy. Carbon dioxide is also made in the process and so the atom economy is not 100%.
   The manufacture of ethanol by the hydration of ethene uses a high temperature and a high pressure. It is a continuous process and the yield is high – the atom economy is 100%. The product needs very little purification. This is not a sustainable process because the main raw material, ethene, is obtained from crude oil.

Acid – a substance that dissolves in water to give a solution with a pH value lower than 7.

Alkali – a substance that dissolves in water to give a solution with a pH value higher than 7.

Alkane – a hydrocarbon molecule containing single bonds only.

Alkene – a hydrocarbon molecule containing a double bond.

Allotropes – different structural forms of the same element, e.g. diamond and graphite are both forms of carbon with different molecular structures.

Alloy – a mixture of two or more metals, or of a metal and a non-metal.

Anode – the positive electrode.

Atom – the smallest part of an element that can enter into chemical reactions.

Atom economy – a measure of how much of the reactants are converted into useful product.

Atomic number – the number of protons in an atom.

Base – a substance that will neutralise an acid.

Batch process – a process in which chemicals are put into a container, the reaction takes place, and the products are removed before a new reaction is started.

Catalyst – a substance that is used to speed up a chemical reaction without being chemically changed at the end of the reaction.

Cathode – the negative electrode.

Compound – a substance consisting of two or more atoms chemically combined together by ionic or covalent bonds.

Concentration – a measure of the amount of substance dissolved in a solution.

Contact process – used to make sulfuric acid.

Continuous process – a process that does not stop; reactants are fed in at one end and products are removed at the other end at the same time.

Covalent bond – a bond between two atoms formed by sharing a pair of electrons.

Cracking – a process used to break up large hydrocarbon molecules into smaller, more useful molecules.

Decomposition – the breaking down of a substance into simpler substances, e.g. using heat or electricity.

Dilute – to reduce the concentration of a substance by adding water.

Electrode – the conducting rod or plate (usually metal or graphite) that allows electric current to enter and leave an electrolysis cell.

Electrolysis – the breaking down of a liquid ionic substance using electricity.

Electrolyte – an aqueous or molten substance that contains free-moving ions and is therefore able to conduct electricity.

Electron – a negatively charged subatomic particle that orbits the nucleus of an atom; relative mass 0.0005.

Element – a substance that consists of only one type of atom.

Empirical formula – gives the simplest whole number ratio of each type of atom in a compound.

Emulsion – a mixture of one liquid finely dispersed in another liquid.

Endothermic – a reaction in which energy is taken in.

Equilibrium – the state in which a chemical reaction proceeds at the same rate as its reverse reaction. (The quantities of reactants and products stay constant.)

Glossary

Eutrophication – the excessive growth and decay of aquatic plants, e.g. algae, due to increased levels of nutrients in the water (often caused by fertilisers or untreated sewage), which results in oxygen levels dropping so that fish and other animal populations eventually die out.

Exothermic – a reaction in which energy is given out.

Fermentation – the process by which yeast converts sugars to alcohol and carbon dioxide through anaerobic respiration.

Fossil fuel – a substance that is burned to release heat energy; formed from the remains of plants and animals over millions of years.

Fraction – a mixture of hydrocarbons with similar boiling temperatures that separated during distillation.

Fractional distillation – a process that uses distillation to separate a mixture such as crude oil into groups of substances with similar boiling points.

Galvanising – coating steel with zinc to prevent corrosion.

Group – a vertical column of elements in the Periodic Table.

Haber process – process used to make ammonia.

Halide – a compound containing a metal and a halogen.

Halogens – elements in Group 7 of the Periodic Table.

Hydrocarbon – a molecule containing hydrogen and carbon only.

Hydrophilic – water loving.

Hydrophobic – water hating.

Immiscible – liquids that do not mix; they form two distinct separate layers.

Insoluble – a substance that is unable to dissolve in a solvent.

Ion – a positively or negatively charged particle formed when an atom gains or loses one or more electron(s).

Ionic bond – the bond formed when electrons are transferred between a metal atom and a non-metal atom, creating charged ions that are held together by forces of attraction.

Isotopes – atoms of the same element that contain different numbers of neutrons.

Limiting reactant – the reactant that gets used up first in a reaction.

Mass number – the total number of protons and neutrons in an atom.

Molar mass – the mass of 1 mole in grams; the relative formula mass in grams.

Mole – the amount of substance that contains 6×10^{23} specified particles.

Molecule – two or more atoms bonded together.

Nanochemistry – the study of materials that have a very small size, in the order of 1–100nm; one nanometre is one billionth of a metre and can be written as 1nm or 1×10^{-9}m.

Neutron – a particle found in the nucleus of atoms; it has no charge; relative mass 1.

Non-renewable – no more can be made in an acceptable time.

Nucleus – the core of an atom, made up of protons and neutrons (except hydrogen, which contains a single proton).

Oxidation – a reaction involving the gain of oxygen or the loss of electrons.

Period – a horizontal row of elements in the Periodic Table.

Periodic Table – a list of the elements arranged to show the trends and patterns in their properties.

Pollutant – a chemical produced by human activity that can harm the environment and organisms.

Polymerisation – the reaction of many monomer molecules joining together to make one large polymer molecule.

Precipitate – an insoluble solid formed during a reaction involving solutions.

Precipitation – the formation of an insoluble solid (a precipitate) when two solutions containing ions are mixed together.

Product – a substance produced in a reaction.

Proton – a positively charged particle found in the nucleus of an atom; relative mass 1.

Radical – a highly reactive atom.

Reactant – a starting material in a reaction.

Redox reaction – a reaction that involves both reduction and oxidation.

Reduction – a reaction involving the loss of oxygen or the gain of electrons.

Relative atomic mass (A_r) – the mass of an atom compared to one-twelfth of the mass of a carbon-12 atom.

Relative formula mass (M_r) – the sum of the atomic masses of all the atoms in a molecule.

Reversible reaction – a reaction in which the products can react to reform the original reactants (equilibrium).

Salt – the product of a chemical reaction between a base and an acid.

Saponification – the process used to make soap by reacting vegetable oil with hot sodium hydroxide.

Saturated – a compound in which all carbon–carbon bonds are single bonds.

Simple distillation – a process used to separate liquids by evaporation followed by condensation to produce a pure liquid.

Soluble – a property that means a substance can dissolve in a solvent.

Solute – a substance that gets dissolved by a solvent.

Solution – the mixture formed when a solute dissolves in a solvent.

Solvent – a liquid that can dissolve another substance to produce a solution.

Synthetic – made or manufactured; not natural.

Titration – a method used to find the concentration of an acid or an alkali.

Titre – the volume of acid delivered by a burette to neutralise an alkali (or vice versa).

Universal indicator – a mixture of pH indicators, which produces a range of colours according to pH and can therefore be used to measure the pH of a solution.

Unsaturated – a compound in which at least one carbon–carbon bond is a double bond.

Yield – the amount of product obtained, e.g. from a crop or a chemical reaction.

HT **Intermolecular** – describes an interaction between one molecule and another.

Spectator ion – an ion not involved in a precipitation reaction.

Subduction – a plate boundary where one tectonic plate is forced below the other and the rock melts into the magma.

Volatile – describes a substance that evaporates quickly.

Notes

Notes

Key

relative atomic mass **atomic symbol** name atomic (proton) number

1	2											3	4	5	6	7	0
							1 **H** hydrogen 1										4 **He** helium 2
7 **Li** lithium 3	9 **Be** beryllium 4											11 **B** boron 5	12 **C** carbon 6	14 **N** nitrogen 7	16 **O** oxygen 8	19 **F** fluorine 9	20 **Ne** neon 10
23 **Na** sodium 11	24 **Mg** magnesium 12											27 **Al** aluminium 13	28 **Si** silicon 14	31 **P** phosphorus 15	32 **S** sulfur 16	35.5 **Cl** chlorine 17	40 **Ar** argon 18
39 **K** potassium 19	40 **Ca** calcium 20	45 **Sc** scandium 21	48 **Ti** titanium 22	51 **V** vanadium 23	52 **Cr** chromium 24	55 **Mn** manganese 25	56 **Fe** iron 26	59 **Co** cobalt 27	59 **Ni** nickel 28	63.5 **Cu** copper 29	65 **Zn** zinc 30	70 **Ga** gallium 31	73 **Ge** germanium 32	75 **As** arsenic 33	79 **Se** selenium 34	80 **Br** bromine 35	84 **Kr** krypton 36
85 **Rb** rubidium 37	88 **Sr** strontium 38	89 **Y** yttrium 39	91 **Zr** zirconium 40	93 **Nb** niobium 41	96 **Mo** molybdenum 42	[98] **Tc** technetium 43	101 **Ru** ruthenium 44	103 **Rh** rhodium 45	106 **Pd** palladium 46	108 **Ag** silver 47	112 **Cd** cadmium 48	115 **In** indium 49	119 **Sn** tin 50	122 **Sb** antimony 51	128 **Te** tellurium 52	127 **I** iodine 53	131 **Xe** xenon 54
133 **Cs** caesium 55	137 **Ba** barium 56	139 **La*** lanthanum 57	178 **Hf** hafnium 72	181 **Ta** tantalum 73	184 **W** tungsten 74	186 **Re** rhenium 75	190 **Os** osmium 76	192 **Ir** iridium 77	195 **Pt** platinum 78	197 **Au** gold 79	201 **Hg** mercury 80	204 **Tl** thallium 81	207 **Pb** lead 82	209 **Bi** bismuth 83	[209] **Po** polonium 84	[210] **At** astatine 85	[222] **Rn** radon 86
[223] **Fr** francium 87	[226] **Ra** radium 88	[227] **Ac*** actinium 89	[261] **Rf** rutherfordium 104	[262] **Db** dubnium 105	[266] **Sg** seaborgium 106	[264] **Bh** bohrium 107	[277] **Hs** hassium 108	[268] **Mt** meitnerium 109	[271] **Ds** darmstadtium 110	[272] **Rg** roentgenium 111	Elements with atomic numbers 112–116 have been reported but not fully authenticated						

*The lanthanoids (atomic numbers 58–71) and the actinoids (atomic numbers 90–103) have been omitted.

Index